AF342053

Offshore Wind Turbines

Offshore Wind Turbines

Reliability, availability and maintenance

Peter Tavner

The Institution of Engineering and Technology

Published by The Institution of Engineering and Technology, London, United Kingdom

The Institution of Engineering and Technology is registered as a Charity in England & Wales (no. 211014) and Scotland (no. SC038698).

The Institution of Engineering and Technology
Michael Faraday House
Six Hills Way, Stevenage
Herts, SG1 2AY, United Kingdom

www.theiet.org

British Library Cataloguing in Publication Data
A catalogue record for this product is available from the British Library

ISBN 978-1-84919-229-3 (hardback)
ISBN 978-1-84919-230-9 (PDF)

Typeset in India by MPS Limited
Printed in the UK by CPI Group (UK) Ltd, Croydon, CR0 4YY

This book is dedicated to
Sarah and Charles.

Behold, the sea itself
And on its limitless, heaving breast, the ships;
See where their white sails, bellying in the wind,
Speckle the green and blue sea.

Walt Whitman, put to music in the Sea Symphony
by Ralph Vaughan Williams.

Contents

Preface

The development of offshore wind power has become a pressing modern energy issue in which the United Kingdom is taking a major part, driven by the need to find new electrical power sources, avoiding the use of fossil fuels, in the knowledge of the extensive wind resource available around our islands and the fact that the environmental impact of offshore wind farms is likely to be low.

However, there are major problems to solve if offshore wind power is to be realised and these revolve around the need to capture this energy at a cost per megawatt hour competitive with other practicable sources. This will depend upon the reliability, availability and longevity of the wind turbines, which make up these offshore wind farms. The cost-effectiveness of the maintenance needed to achieve that availability and longevity is essential to improve offshore wind life-cycle costs and the future of this emerging industry.

This book intends to address these issues head-on and demonstrates clearly to manufacturers, developers and operators the facts and figures of wind turbine operation and maintenance in the inclement offshore environment, recommending how maintenance should be done to achieve low life-cycle costs.

The author has been working on this problem for 10 years, but his main technical experience was in the conventional fossil- and nuclear-fired electricity supply industry operating and manufacturing power equipment, from which many lessons can be learnt about wind industry through-life costs. However, modern fossil- and nuclear-fired power stations are in effect purpose-designed, well-housed, power factories, manned 24 hours a day 7 days a week, whose effectiveness has been demonstrated over the past 80 years. The author also had an early naval training and knows from ship operations the role that good design, manufacture and maintenance must play in keeping a ship operational on the high seas also assisted by the fact that ships are manned 24/7. The efficacy of our maritime trade over the last 100 years shows how this can be achieved. The offshore oil and gas industry, particularly in the North Sea, where many offshore wind assets are and will be installed, has also learnt how to install, maintain and operate effective offshore engineering structures over the past 40 years, including some lessons about operating at reduced manning levels.

Offshore wind technology has some similarities to all of the above but consists of unmanned, robotic power units operating 24/7, controlled from remote onshore control rooms where manning levels are low. The engineering issues facing us as we build, maintain and keep these wind power stations at high degrees of operational readiness with those low manning levels present fascinating challenges, which our power station, marine and offshore oil and gas experiences will assist to overcome. However, the offshore wind industry also needs innovation, new technology, good manufacture and excellent management to become successful.

Andrew Garrad, the co-founder of the UK wind consultancy now called GL Garrad Hassan, has said 'that for a long time the mantra of the wind turbine industry has been bigger and bigger but now it has moved to better and better and this change marks a change in the areas of innovation' (Jamieson, 2011).

I hope that this book, written from a UK perspective and based upon our own research at Durham, will help you to achieve that for the future.

Peter Tavner
Durham University

Acknowledgements

I am indebted not only to a number of colleagues who have contributed to this book, most particularly my PhD students Michael Wilkinson, Fabio Spinato, Chris Crabtree, Chen Bin Di, Mahmout Zaggout and Donatella Zappala, but also to undergraduate and research students and post-doctoral research workers including Hooman Arabian-Hoseynabadi, Lucy Collingwood, Sinisa Djurovic, Yanhui Feng, Rosa Gindele, Andrew Higgins, Mark Knowles, Ting Lei, Luke Longley, Yingning Qiu, Paul Richardson, Sajjad Tohidi, Wenjuan Wang, Xiaoyan Wang, Matthew Whittle, Jianping Xiang and Wenxian Yang. I am also indebted to academic colleagues Rob Dominy, Simon Hogg, Hui Long, Li Ran and William Song in Durham; Geoff Dutton, Bill Leithead, Sandy Smith and Simon Watson, members of the Supergen Wind Consortium; Berthold Hahn, Stefan Faulstich, Joachim Peinke, Gerard van Bussel and other European Academy of Wind Energy colleagues who have opened my eyes towards the knowledge developed in wind power throughout Europe. I would particularly also like to thank Peter Jamieson, formerly of GL Garrad Hassan, but now of Strathclyde University, for his long knowledge and experience of the wind industry and for his recent book on innovation in wind turbine design (Jamieson, 2011), which is so important in the area of improving through-life performance. May I also mention Professor Jurgen Schmid of Kassel University, a founder of the European Academy of Wind Energy, who started the investigation into wind turbine reliability with the first book on the subject in 1991 (Schmid & Klein, 1991). I also acknowledge the assistance in preparing Appendix 2 of ReliaWind Partners, in particular GL Garrad Hassan, and advice from E.ON Climate and Renewables.

I would also like to express appreciation for the research funding that has made this book possible, to the UK Engineering and Physical Science Research Council for the Supergen Wind Phases 1 and 2 funding and to the European Union for the Framework Programme 7 ReliaWind Consortium funding. Finally, I would like to thank colleagues in a number of industrial organisations who have assisted by providing data or photographs, including Alnmaritec, ABB Drives, Alstom Wind Power, Clipper Wind Marine, Converteam GL Garrad Hassan, Hansen Transmissions MTS, National Renewable Energy Centre (UK) and National Renewable Energy Laboratory (USA), Siemens Wind Power and Wind Cats. Chris Orton of Durham University carefully prepared the diagrams.

Nomenclature

Symbol	Explanation
A	For a WT class, this designates the category for higher turbulence characteristics
B	For a WT class, this designates the category for medium turbulence characteristics
C	For a WT class, this designates the category for lower turbulence characteristics
A	Availability, $A = MTBF/(MTBF + MTTR)$
$A(t)$	Availability function of a population of sub-assemblies as a function of time
Acc	Acceleration factor for accelerated life testing
AEP	Annual energy production (MWh)
C	Capacity factor (%)
CoE	Cost of energy (£/MWh)
$F(t)$	Failure intensity, can be represented by a PLP or Weibull function
F or F^{-1}	Forward or backward Fast Fourier Transform
FCR	Fixed charge rate for interest (%)
η	Efficiency
H_s	Wave height for sea state
ICC	Initial capital cost (£)
I	Drive train inertia (kg m^2)
I	Turbulence intensity, defined by IEC 61400 Part 1, σ/u
I_{char}	Turbulence characteristic, defined by IEC 61400 Part 1
I_{ref}	Expected value of turbulence intensity at u_{ref} 15 m/s
k	Constant in power balance equations
ku_n	Turbulence coefficient at a wind speed u of n (m/s)

(*Continues*)

(*Continued*)

Symbol	Explanation
$\lambda(t)$	Instantaneous hazard function for a sub-assembly or machine, failures/sub-assembly/year
λ	Failure rate of a sub-assembly or machine varying with time, failures/sub-assembly/year
N	Speed of a machine rotor (rev/min)
n	Number of years
P	Power (Watt)
P_{det}	Probability of detection of a fault
p	Integer number of pole pairs
Q	Heat flow (Watt/m^2)
R	Resistance (Ohm)
$R(t)$	Reliability or survivor function of a population of sub-assemblies as a function of time (failures/machine/year)
r	Discount rate (%)
S	Specific energy yield (MWh/m^2/yr)
σ	Wind speed standard deviation
T	Torque (Nm)
T	Temperature (°C)
ΔT	Temperature rise (°C)
T	Period of a wave (second)
u	Wind speed (m/s, mile/hr, knot)
θ	MTBF of a sub-assembly, $\theta = 1/\lambda$ (hours)
V_{ref}	Mean wind speed at WT hub height (m/s)
V	Rms voltage (V)
W	Work done in a WT drive train
ω	Angular frequency (rad/s)

Abbreviations

Symbol	Explanation
AEP	Annualised energy production
AIP	Artemis Innovative Power
ALT	Accelerated life testing
AM	Asset management
AMSAA	Army Materiel Systems Analysis Activity
BDFIG	Brushless doubly fed induction generator
BMS	Blade Monitoring System
BOP	Balance of Plant
CAPEX	Capital expenditure
CBM	Condition-based maintenance
CMS	Condition Monitoring System
CoE	Cost of energy
DCS	Distributed Control System
DDPMG	Direct drive permanent magnet synchronous generator
DDT	Digital Drive Technology (AIP)
DDWRSGE	Direct drive wound rotor synchronous generator and exciter
DE	Drive end of generator or gearbox
DFIG	Doubly fed induction generator
EAWE	European Academy of Wind Energy
EFC	Emergency feather control
EPRI	Electric Power Research Institute, USA
EWEA	European Wind Energy Association
FBG	Fibre Bragg Grating
FCR	Fixed charge rate, interest rate on borrowed money
FFT	Fast Fourier Transform

(Continues)

(*Continued*)

Symbol	Explanation
FM	Field maintenance
FMEA	Failure Modes and Effects Analysis
FMECA	Failure Modes, Effects and Criticality Analysis
FSV	Field support vessel
HAWT	Horizontal axis wind turbine
HM	Health monitoring
HPP	Homogeneous Poisson process
HSS	Gearbox high-speed shaft
HV	High voltage
ICS	Integrated Control System
IEC	International Electrotechnical Commission
IEEE	Institute of Electrical and Electronic Engineers
IET	Institution of Engineering and Technology (former IEE)
IM	Information management
IMS	Gearbox intermediate shaft
IP	Intellectual property
LCC	Life cycle costing
LSS	Gearbox low-speed shaft
LV	Low voltage
LWK	Landwirtschaftskammer Schleswig-Holstein database for Germany
MCA	Marine and Coastguard Agency
MIL-HDBK	US Reliability Military Handbook
MM	Maintenance management
MTBF	Mean time between failures
MTTR	Mean time to repair
MV	Medium voltage
NDE	Non-drive end of generator or gearbox
NHPP	Normal homogeneous Poisson process
NPRD	Non-electronic Parts Reliability Data
O&M	Operations and maintenance
OEM	Original equipment manufacturer

(*Continued*)

Symbol	Explanation
OFGEM	Office of Gas and Electricity Markets
OFTO	Offshore Transmission Operator
OM	Operations management
OPEX	Operational expense
OREDA	Offshore Reliability Data
OWT	Offshore wind turbine
PLC	Programmable logic controller
PLP	Power law process
PMG1G	Permanent magnet synchronous generator with 1-stage gearbox
PMSG	Permanent magnet synchronous generator
PSD	Power spectral density
RBD	Reliability block diagram
RMP	Reliability modelling and prediction
RNA	Rotor nacelle assembly
RPN	Risk Priority Number
SCIG	Squirrel cage induction generator
TBF	Time between failures
TTF	Time to failure
TTT	Total time on test
VAWT	Vertical axis wind turbine
WF	Wind farm
WMEP	Wissenschaftlichen Mess- und Evaluierungsprogramm database
WRIG	Wound rotor induction generator
WRIGE	Wound rotor induction generator and exciter
WRSGE	Wound rotor synchronous generator and exciter
WSD	Windstats database for Germany
WSDK	Windstats database for Denmark
WT	Wind turbine
WTCMTR	Wind turbine condition monitoring test rig

Chapter 1
Overview of offshore wind development

1.1   Development of wind power

The human development of rotating machine wind power started more than 2000 years ago at various locations around the globe but particularly in Iran and China, see Chapter 10, Appendix 1.

However, the technology of wind turbines (WTs) for generating electricity dates back to the end of the nineteenth century to three historic WTs: a horizontal-axis wind turbine (HAWT) in the United States in 1883 (the Brush turbine), a vertical-axis wind turbine (VAWT) in Scotland in 1887 (the Blyth turbine) and an HAWT in Denmark in 1887 (the la Cour turbine).

Large electric power WTs >100 kW, <1 MW, were envisaged and built in Germany, Russia and the United States in the 1930s and 1940s. However, the modern large WT developments date back to work in Europe and the United States, later stimulated by European Union (EU) and US Department of Energy experimental programmes in the 1970s to 1980s, following the oil price rises after the 1973 Yom Kippur War between Egypt, Syria and Israel. A detailed description of the WT development is given with photographs in Appendix 1, but the key large WT projects of the last 80 years are listed in Table 1.1 and their evolution has been profoundly influenced by reliability and availability issues.

This design evolution, with competing VAWT or HAWT, two or three blades, upwind or downwind and geared or direct drive configurations, has affected subsequent developments, which is interesting as the reliability of many of these early onshore WT prototypes was extremely poor.

The machines at Grandpa's Knob (the United States), Orkney (the United Kingdom) and Growian (Germany) only operated for some hundreds of hours, suffering catastrophic failures in the turbine hub or blades. But the Gedser machine ran for 11 years without extensive maintenance; this successful configuration, built upon as the Danish Concept, has come to dominate the development of modern WTs.

From these small beginnings, modern wind electrical power generation has expanded rapidly to the present day, as represented by Figure 1.1, showing the world installed capacity.

The recording of WT reliability started in Europe in 1985 [1], with the growth of the German and Danish wind industry, and in the United States in 1987,

Table 1.1   WT development worldwide 1931–2011

Year	Location	Type	Power (MW)	Rotor diameter (Xm)	Tower height (m)	Blade no	Drive	Pitch	Speed	Comment
1931	WIMIE-3D, Yalta, USSR	Upwind HAWT	0.10		30	3	Geared drive	Adjustable blade flaps	Variable speed	Connected to 6.3 kV distribution system; 32% capacity factor; post-mill with the whole structure rotate along track; early large 3-blade machine
1941	Grandpa's Knob, Vermont, USA	Downwind HAWT	1.25	57	40	2	Geared drive	Pitch controlled, stall regulated	Fixed speed	Grid connected
1951	John Brown Engineering, Orkney, UK	Upwind HAWT	0.10	18		3	Geared drive	Full-span pitch regulated	Fixed speed	Grid connected
1956	Station d'Etude de l'Energie du Vent, Nogent-le-Roi, France	Downwind HAWT	0.80			3	Geared drive	Full-span pitch regulated	Variable speed	Grid connected
1956	Johannes Juul, Gedser, Denmark	Upwind HAWT	0.20	24		3	Geared drive	Fixed pitch, stall regulated; aerodynamic tip brakes on rotor blades automatically in over-speed	Fixed speed	The so-called Gedser Mill, defining the Danish 3-blade concept

1979	Nibe, Denmark	Upwind HAWT	0.63			3	Geared drive	Fixed pitch, stall regulated	Fixed speed	Danish Concept
1980	Nibe, Denmark	Upwind HAWT	0.63			3	Geared drive	Full-span pitch regulated	Fixed speed	
1981	Boeing, MOD2, USA	Downwind HAWT	2.50	91	60	2	Geared drive	Full-span pitch regulated	Variable speed	
1983	Große Wind-energieanlage (Growian), Germany	Downwind HAWT	3.00	100	100	2	Geared drive	Full-span pitch regulated	Variable speed	Grid connected with fully rated cycloconverter
1985	Wind Energy Group, LS1, Orkney, UK	Upwind HAWT	3.00	60		2	Geared drive	Adjustable tip-flap pitch regulated	Variable speed	Grid connected with fully rated converter
2007	Enercon E126, Cuxhaven, Germany	Upwind HAWT	7.58	126	135	3	Direct drive	Full-span pitch regulated	Variable speed	Grid connected with fully rated converter

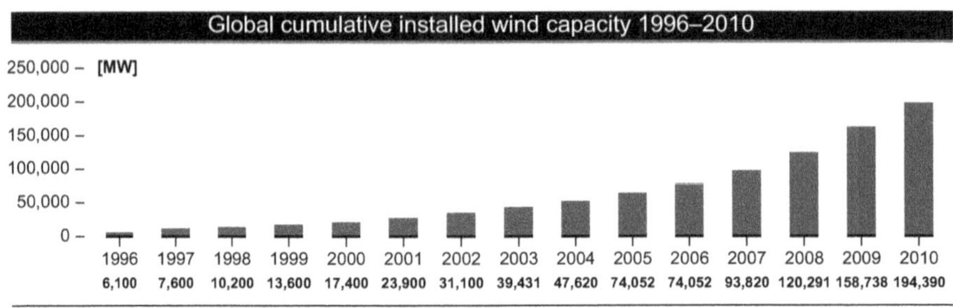

*Figure 1.1   Development of annually installed wind capacity worldwide
1996–2010*

following the growth of US wind farms after 1973. Various reports summarising WT reliability have been published including those given in References 2 and 3. Work in the Netherlands in the 1990s [4], when offshore wind farms were contemplated in the North Sea off the Dutch Coast, lead to concerns about the influence of maintenance access to the WTs and a wider consideration not only of reliability but also of maintenance and the need to achieve a high WT availability. This would lead to a low cost of energy for wind power so that it could compete against low-cost fossil fuels.

Energy production from onshore turbines of larger sizes >1 MW based on the Danish Concept is now achieving operational availabilities of >98% and mean time between failures (MTBFs) of >7000 hours, which is a failure rate of just over 1 failure(s)/turbine/year, where a failure could be described as a stoppage with a duration of 24 hours. The results of early recording of WT failures are summarised in Figure 1.2. The details of these reliability developments will be dealt with in Chapter 2.

Figure 1.2 taken from Reference 2 shows the steady improvement in onshore WT reliability from 1987 to 2005, taken from various public domain sources, in comparison with other grid-connected and distributed generation sources. However, reliability still needs further improvement, and this situation will be substantially affected by deployment offshore.

1.2   Large wind farms

Deployment of WTs in large wind farms has been a feature of modern wind power since the 1980s as we try to harness the geographical extent of the distributed wind resource. The California wind farms built in the 1970s and 1980s (see Figure 1.3) were established with large numbers of relatively small WTs, ≤100 kW arranged in arrays of more than 100 WTs.

An advantage of an extensive wind farm is that the combined electrical resource will be substantial, justifying the cost of grid connection and considerable

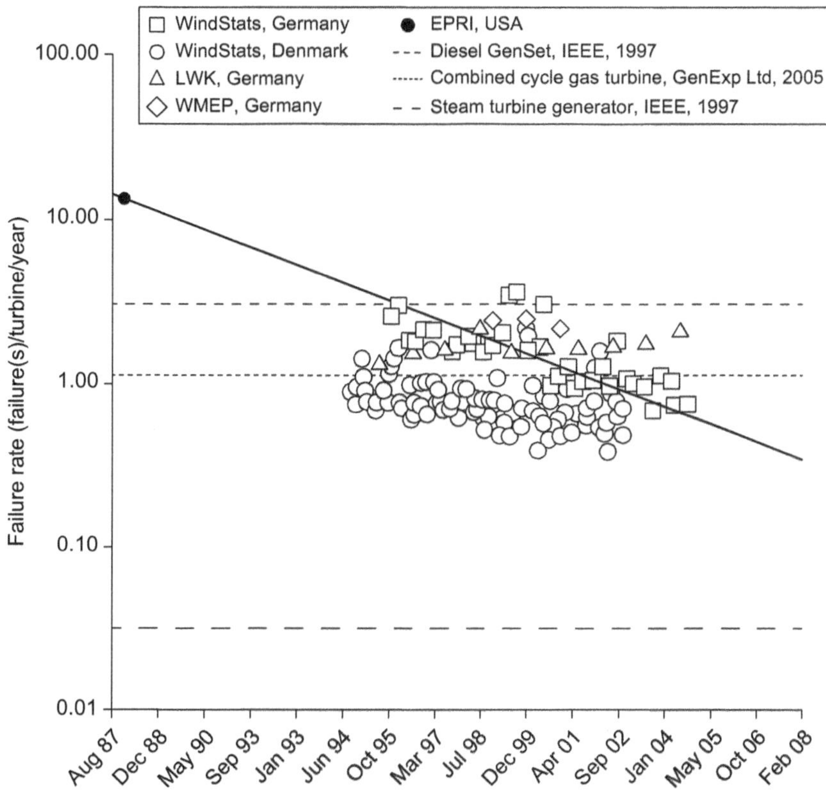

Figure 1.2   *Gross failure rate trends for onshore WTs over the period 1987–2008 [Source: [2]]*

maintenance benefits accrue for a large wind farm because personnel, tools, parts and facilities can be concentrated at or close to the WT farm site. It is currently not possible to tell whether the increasing reliability of WTs, shown in Figure 1.2, can be partially ascribed to their deployment in larger wind farms, although it is likely that this is a contributory factor.

The principle disadvantage of the large onshore wind farm is its visual impact, and this is particularly important in crowded countries, such as the United Kingdom, where citizens put space, amenity and visual impact high on the agenda during any wind farm approval process; In general, while large wind farms have been established in the United States, Spain and northern Germany, they are not common in the United Kingdom where the planning process has militated against the concentration process; therefore UK onshore wind farms have generally ranged from only 1 to 30 WTs. However, the largest onshore wind farm currently operating in the United Kingdom, opened in 2010, is at Whitelee, close to Glasgow (Figure 1.4), which has 140 Siemens 2.3 MW HAWTs.

Figure 1.3   Example of a large wind farm of >100 WTs in California in the early 1980s

Figure 1.4   The largest wind farm in the United Kingdom at Whitelee near Glasgow with 140 Siemens 2.3 MW HAWTs

1.3   First offshore developments

The first offshore wind farm was deployed in Denmark in 1991 at Vindeby with 11 WTs in sheltered, non-tidal Baltic waters close to Fyn island. A small offshore wind farm was installed in the tidal waters of the North Sea close inshore at Blyth, Northumberland, the United Kingdom (2 WTs) in 2001 (see Figure 1.5).

The large capital investment required for offshore installation has subsequently encouraged developers to increase the extent of later offshore wind farms. The first

*Figure 1.5   The first offshore wind farm in UK at Blyth, 2 Vestas V66 HAWTs
[Source: AMEC Border Wind]*

substantial offshore wind farm was installed at Middelgrunden near Copenhagen in Denmark in 2000, with 20 Siemens SWT1.0/54 WTs (see Figure 1.6).

1.4   Offshore wind in Northern Europe

1.4.1   Overview

A summary of current and planned offshore wind farms in Northern Europe in Table 1.2 clearly shows the smaller earlier wind farms in Denmark and the United Kingdom with an expanding size as the years advance with further developments in Germany, the Netherlands and Sweden. The cumulative power generation capacity of the wind farms listed in Table 1.2 is 5.3 GW. Table 1.2 is further summarised in Figure 1.6, which shows the increasing offshore wind farm sizes in Northern Europe. Research in the Netherlands on their offshore programme has been reported in Reference 5.

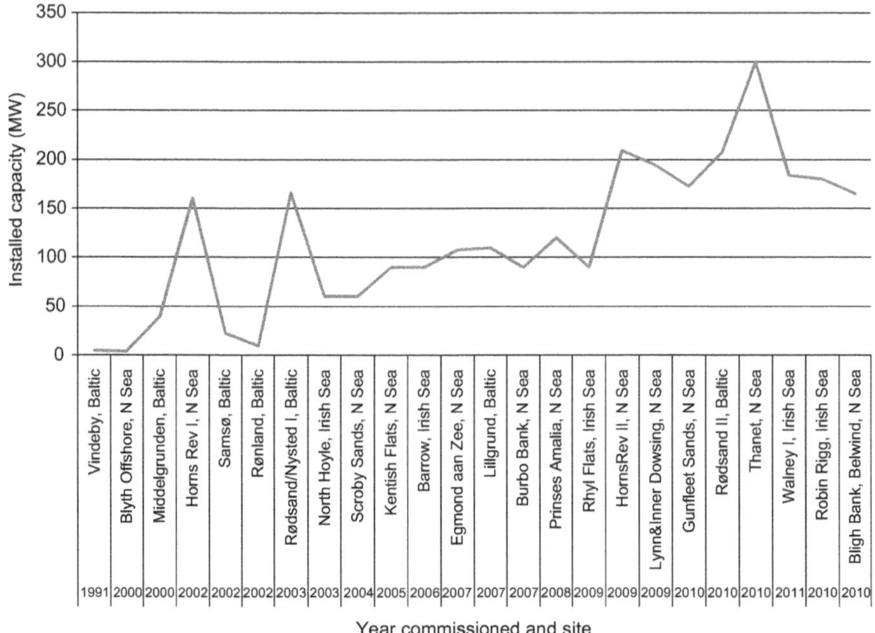

Figure 1.6   North European growth in offshore wind farm size, 1991–2010

1.4.2   Baltic Sea

The Baltic Sea has non-tidal but windy conditions with potential ice and wave hazards. The first large offshore wind farm in the Baltic Sea was installed in 2000 at Middelgrunden (20 WTs) close to Copenhagen in Denmark (see Figure 1.7). This process has accelerated rapidly since the Middelgrunden installation with a number of offshore wind farms being installed including Nysted, Denmark (72 WTs); Lillgrund, Sweden (48 WTs) (Figure 1.8) and Rodsand, Denmark (90 WTs).

1.4.3   UK waters

After the UK Blyth installation, a process of licensing of UK offshore wind farm sites was initiated from the Crown Estate in three rounds. Round 1 adopted a cautious approach, with a model of 25 or 30 WTs per wind farm, intended to allow developers, installers and operators to gain experience. This has proved a successful model and its caution can be seen at the centre of Figure 1.6. In Denmark, after accelerating the process in more benign Baltic waters, offshore wind farm size was dramatically increased at Horns Rev 1 (80 WTs) in the North Sea. Operational problems in the first years of operation at Horns Rev, caused essentially by onshore WTs being installed offshore, then lead to a major rethink by WT OEMs (original equipment manufacturers) and wind farm developers of future North Sea designs,

Table 1.2   European offshore wind farms under construction up to 2011

Wind farm	Capacity (MW)	Country	WT no.	Maker	Type	Turbine rating (MW)	Commissioned
Vindeby	4.95	Denmark	11	Siemens		0.45	1991
Blyth Offshore	4	UK, Round 1	2	Vestas	V66	2.0	2000
Middelgrunden	40	Denmark	20	Siemens	SWT-2.0-76	2.0	2000
Hcms Rev I	160	Denmark	80	Vestas	V80	2.0	2002
Samsø	23	Denmark	10	Siemens	SWT-2.3-82	2.3	2002
Rønland	9.2	Denmark	4	Siemens	SWT-2.3-93	2.3	2002
Rødsand/Nysted I	166	Denmark	72	Siemens	SWT-2.3-82	2.3	2003
Frederikshavn	2.3	Denmark	1	Siemens	SWT-2.3-82	2.3	2003
North Hoyle	60	UK, Round 1	30	Vestas	V80	2.0	2003
Scroby Sands	60	UK, Round 1	30	Vestas	V80	2.0	2004
Kentish Flats	90	UK, Round 1	30	Vestas	V90	3.0	2005
Barrow	90	UK, Round 1	30	Vestas	V90	3.0	2006
Egmond aan Zee	108	Netherlands	36	Vestas	V90	3.0	2007
Lillgrund	110	Sweden	48	Siemens	SWT-2.3-93	2.3	2007
Burbo Bank	90	UK, Round 1	25	Siemens	SWT-3.5-107	3.6	2007
Beatrice	10	UK	2	RePower	5M	5.0	2007
Prinses Amalia	120	Netherlands	60	Vestas	V80	2.0	2008
Hywind	2.3	Norway	1	Siemens	SWT-2.3-82	2.3	2009
Rhyl Flats	90	UK, Round 1	25	Siemens	SWT-3.5-107	3.6	2009
Horns Rev II	209	Denmark	91	Siemens	SWT-2.3-92	2.3	2009
Lynn & Inner Dowsing	194	UK, Round 1	54	Siemens	SWT-3.5-107	3.6	2009
Alpha Ventus	60	Germany	12	RePower & Areva	5M & M 5000	5.0	2009
Gunfleet Sands	173	UK, Round 1	48	Siemens	SWT-3.6-107	3.6	2010
Rødsand II	207	Denmark	90	Siemens	SWT-2.3-93	2.3	2010
Thanet	300	UK, Round 2	100	Vestas	V90	3.0	2010

(*Continues*)

Table 1.2   (*Continued*)

Wind farm	Capacity (MW)	Country	WT no.	Maker	Type	Turbine rating (MW)	Commissioned
Walney I	184	UK, Round 2	51	Siemens	SWT-3.6-107	3.6	2011
Robin Rigg	180	UK, Round 2	60	Vestas	V90	3.0	2010
Baltic I	48	Denmark	21	Siemens	SWT-2.3-93	2.3	2010
Bligh Bank, Belwind	165	Belgium	55	Vestas	V90	3.0	2010
Greater Gabbard	504	UK, Round 2	140	Siemens	SWT-3.6-107	3.6	
London Array	630	UK, Round 2	175	Siemens	SWT-3.6-120	3.6	
Sheringham Shoal	317	UK, Round 2	88	Siemens	SWT-3.6-107	3.6	
Anholt	400	Denmark	111	Siemens	SWT-3.6-120	3.6	
Pori	2.3	Finland	1	Siemens	SWT-2.3-101	2.3	
Walney II	183	UK, Round 2	51	Siemens	SWT-3.6-120	3.6	
Borkum Riffgat	108	Denmark	30	Siemens	SWT-3.6-107	3.6	
Baltic II	288	Denmark	80	Siemens	SWT-3.6-120	3.6	
Dan Tysk	288	Denmark	80	Siemens	SWT-3.6-120	3.6	
TOTAL	**5675**						

Figure 1.7   First Baltic large offshore wind farm, Middelgrunden, Copenhagen, 20 Siemens SWT1.0 HAWTs

Figure 1.8   A large Swedish offshore wind farm at Lillgrund, 48 WTs

slowing down development. In the subsequent UK Round 2, the size of planned wind farms has increased to >50 WTs but has been slow to develop. However, early operational success with the smaller UK Round 1 sites, where the severe problems at Horns Rev were largely avoided, even though some sites used the same WTs, has encouraged developers. Therefore, the installation of Round 2 wind farms is now accelerating, with the first of these operational in 2011 at Thanet (100 WTs). Meanwhile Dutch, Belgian and Danish developers have similarly accelerated their large North Sea installations at Prinses Amalia (60 WTs), Belwind (55 WTs) and Horns Rev II (91 WTs).

In the United Kingdom, Round 3 is considering much larger arrays of 5–600 WTs, but these wind farms are still in the planning stage.

1.5   Offshore wind rest of the world

1.5.1   The United States

No offshore wind farms have yet been built in the United States, but considerable resource measurement and development is underway to consider offshore wind farm sites on the eastern seaboard.

1.5.2   Asia

China has started the development of an offshore wind industry and has so far installed three small wind farms as shown in Table 1.3. Work was initiated cautiously with 1 WT in Bohai Bay in 2007 and at an inter-tidal wind farm at Rudong (16 WTs). A larger wind farm is under construction at Donghai Bridge, Shanghai (34 WTs), and Figure 1.9 shows one of these 3 MW turbines being installed.

Table 1.3   China offshore and inter-tidal wind farms

Wind farm	Type	Capacity (MW)	Province	WT no.	OEM and type	Commissioned
Bei Hai	Offshore, connected to offshore oil platform	1.5	Liaoning	1	Goldwind 1.5 MW	2007
Rudong	Inter-tidal, grid connected	30	Jiangsu	16	Various manufacturers	2009
Dong Hai	Offshore, grid connected	102	Shanghai	34	Sinovel SL3000/90	2010

1.6   Offshore wind power terminology and economics

1.6.1   Terminology

The definition of availability for WTs needs to be clarified. Since 2007, an International Electrotechnical Commission working group has been working to produce

Figure 1.9　Installation of a 3 MW Sinovel WT at Dong Hai near Shanghai

a standard IEC 61400-Pt 26 to define WT availability in terms of time and energy output. Until that standard is published, however, there is no internationally agreed definition of availability either in terms of time or energy. However, two availability definitions have been generally adopted in the United Kingdom in reports [6] and are summarised below.

- Technical availability, also known as system availability, is the percentage of time that an individual WT or wind farm is available to generate electricity expressed as a percentage of the theoretical maximum.
- Commercial availability, also known as turbine availability, is the focus of commercial contracts between wind farm owners and WT OEMs to assess the operational performance of a wind farm project. Some commercial contracts may exclude downtime for agreed items, such as requested stops, scheduled repair time, grid faults and severe weather, when WTs cannot operate normally.

For the rest of the book, the term 'availability' refers either to technical availability as defined above lending itself to comparison from project to project.

From the above definitions, it follows that technical availability will always be lower than the commercial availability because there is more alleviation of downtime for the former, and an important issue offshore is that availability, A, is affected by both time and wind speed, u, $A(u, t)$ [7].

In respect of reliability, the following expressions are useful:

$$
\begin{aligned}
&\text{Mean time to failure} \quad &MTTF \\
&\text{Mean time to repair} \quad &MTTR \\
&\text{Logistic delay time} \quad &LDT \\
&\text{Downtime} \quad &MTTR + LDT
\end{aligned}
\tag{1.1}
$$

$$\text{Mean time between failure} \quad MTBF \approx MTTF \tag{1.2}$$

$$MTBF \approx MTTF + MTTR = \frac{1}{\lambda} + \frac{1}{\mu} \tag{1.3}$$

$$MTBF = MTTF + MTTR + LDT \tag{1.4}$$

$$\text{Failure rate, } \lambda = \frac{1}{MTBF} \tag{1.5}$$

$$\text{Repair rate, } \mu = \frac{1}{MTTR} \tag{1.6}$$

$$\text{Commercial availability, } A = \frac{MTBF - MTTR}{MTBF} = 1 - \left(\frac{\lambda}{\mu}\right) \tag{1.7}$$

$$\text{Technical availability, } A = \frac{MTTF}{MTBF} < 1 - \left(\frac{\lambda}{\mu}\right) \tag{1.8}$$

Note that these are all expressed in terms of the variable time, but availability can be expressed in terms of energy production and this will ultimately be more valuable for the operator (Figure 1.10).

Capacity factor and specific energy yield are two commonly used terms describing the productivity of a WT or wind farm. Capacity factor, C, is defined as the percentage of the actual annual energy production E (MWh) over the rated annual energy production, AEP, from a WT or wind farm of rated power output P:

$$C = AEP \times \frac{100}{P \times 8760}\% \tag{1.9}$$

Specific energy yield, S (MWh/m^2/yr), is defined as the AEP of a WT normalised to its swept rotor area, A (m^2):

$$S = \frac{AEP}{A} \tag{1.10}$$

The ratio, R_S, of rated power, P, over the swept rotor area, A, is a fixed value for a specific WT type:

$$R_s = \frac{P}{A} \tag{1.11}$$

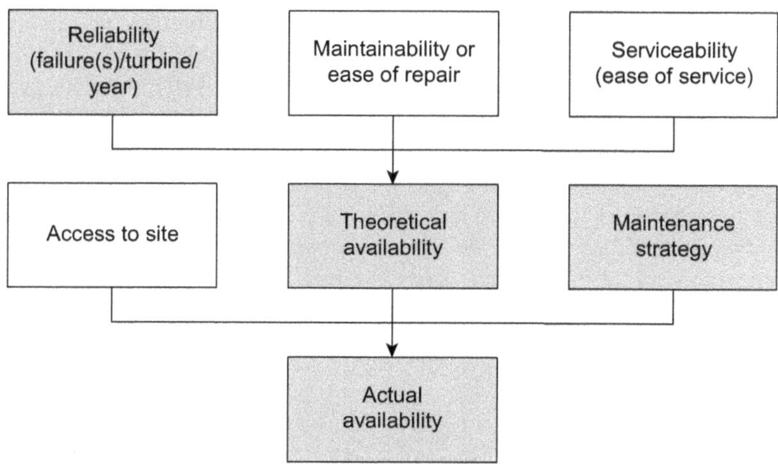

Figure 1.10   Availability as a function of machine properties, access to site accessibility and maintenance strategy [Source: [8]]

or

$$RS = \frac{S}{C \times 8760}$$ (1.12)

For a specific type of WT, the specific energy yield is proportional to the capacity factor:

$$S = R_S \times C \times 8760$$ (1.13)

Therefore, the operational performance of a WT or wind farm can be defined as the percentage of the achieved over the expected C or S.

1.6.2   Cost of installation

Offshore wind power uses large WTs whose capital cost is currently estimated at around £1.2 million/MW, compared to onshore WTs at £0.65 million/MW [6]. Offshore wind turbine (OWT) structures are large; the WT hub for a 3.5 MW offshore machine will be 90 m above the sea surface; the rotor diameter will be of the order of 100 m. Initially the structures will be installed in relatively shallow water depth, 5–20 m, and the weight of each structure will be relatively low, ≈ 400 tonnes, depending on rating. So, in contrast to typical oil and gas onshore structures, the applied vertical load to the foundation is relatively small compared to the wind and wave overturning moments. Therefore, an OWT foundation may account for up to 35% of the installed cost [6]. Therefore, OWT unit capital costs are large and will increase as the wind farms are placed in deeper water.

However, a single OWT design can be mass-produced for use over a whole wind farm or many wind farms, rather than each structure/foundation being individually engineered, as it would be in the oil and gas industry. So capital costs of OWTs will fall progressively with subsequent projects at later times and this has been noted in the Danish, Swedish, the UK, German and Dutch offshore projects.

An interesting comparison can be made between the capital cost for offshore wind in China at Dong Hai Da Qiao compared with UK late Round 1 projects as shown in Figure 1.11. The capital costs of offshore wind in China at £2.15 million/MW are greater than in the United Kingdom at £1.25 million/MW because China is at the very start of its offshore development, whereas the United Kingdom has already learnt some of the lessons. Costs in China will fall as capacity increases.

Further details on costs are given in Reference 9.

1.6.3   Cost of energy

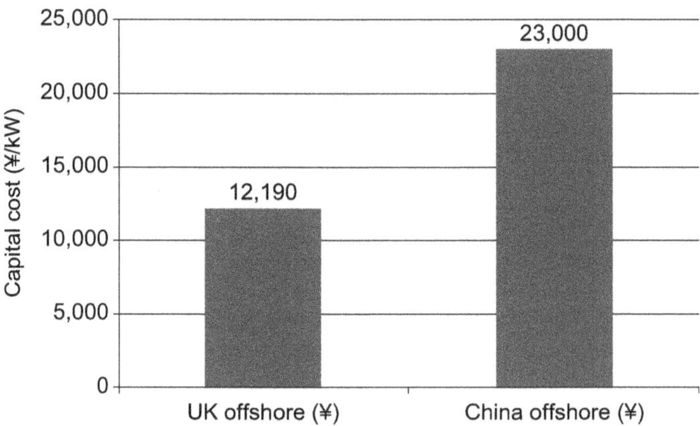

Figure 1.11   Comparison of offshore wind capital cost between the United Kingdom and China

Cost of energy (*CoE*) is commonly used to evaluate the economic performance of different wind farms. This methodology was adopted in a joint report [10] by the International Energy Agency (IEA), the European Organisation for Economic Co-operation and Development (OECD) and US Nuclear Energy Agency (NEA). It compared the cost of different electricity production options. A simplified calculation equation was adopted in the United States to calculate the *CoE* (£/MWh) for a WT system [11]:

$$CoE = \frac{ICC + FCR\ O\&M}{AEP} \tag{1.14}$$

where *ICC* is initial capital cost (£); *FCR* is annual fixed charge rate (%); *AEP* is annual energy production (MWh) and *O&M* is annual O&M (operations and maintenance) cost (£).

The result of this approach is the same as that of levelised electricity generation cost used in Reference 11, where the parameter *FCR* is a function of the discount rate *r* used as follows:

$$FCR = \frac{r}{1 - (1 + r)^n} \qquad (1.15)$$

where $r \neq 0$. The discount rate *r* is the sum of inflation and real interest rates. If inflation is ignored, the discount rate equals the interest rate. For the special case of a discount rate $r = 0$, unlikely in the real world, *FCR* will be *ICC* divided by the economic lifetime of the wind farm in years, currently estimated at $n = 20$ years.

A preliminary estimation of the *CoE* from offshore wind was carried out in Reference 12 on the early UK Round 1 sites. This shows that at that stage the *CoE* for offshore wind in the United Kingdom was about $1.5\times$ that for onshore (see Figure 1.12). It is probable that improvements in λ and μ will have improved these figures.

The UK subsidised *CoE* for offshore wind is therefore estimated from Round 1 at about £69/MWh against £47/MWh for onshore. An interesting comparison (Figure 1.13) can be made with the *CoE* for offshore wind from the Shanghai Donghai Bridge project in China of ¥980/MWh (i.e. ~£91/MWh), on a project installation cost of ¥23,000/kW (i.e. ~£2150/kW) from Chinese sources. Again it should be expected that these *CoE* will fall as experience is gained, the O&M costs fall and the risks associated with the capital investment reduce.

These calculations were made on the basis of the subsidised *CoE*, and recent work has stripped away those benefits showing the true *CoE* for offshore wind around the UK coast to be closer to £140/MWh. Again this will fall as experience is gained and capital costs fall and life is extended, the latter being heavily influenced by the O&M regime surrounding the wind farm. Early studies show clearly that operators who impose a higher quality O&M regime achieve higher availability, lower through-life costs and a lower *CoE*. The relationship between *CoE* and the

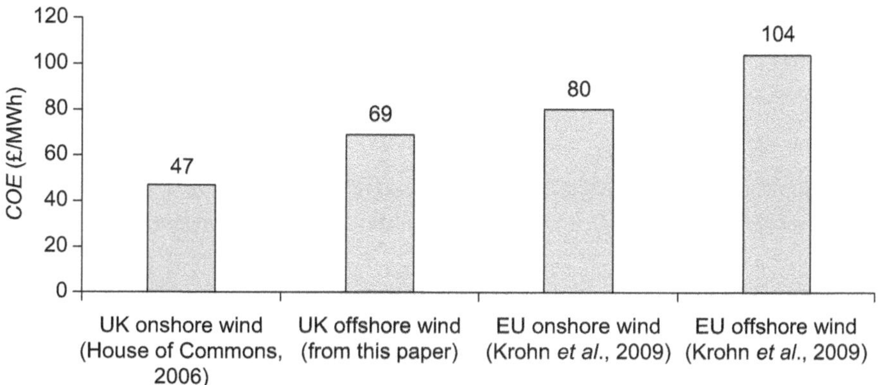

Figure 1.12   Relative CoE for offshore wind in the United Kingdom and Europe [Source: [12]]

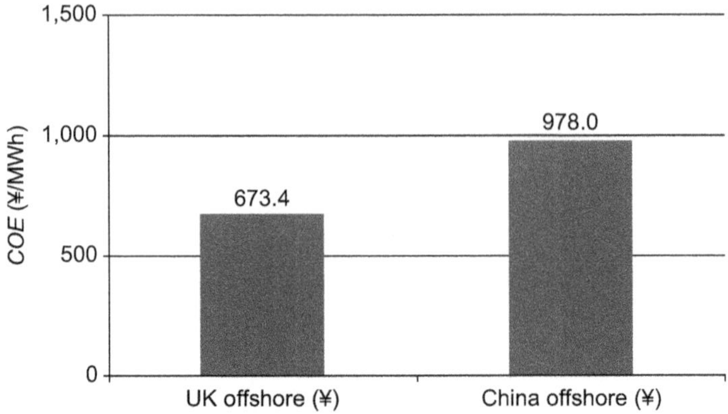

Figure 1.13   Comparison of offshore wind power CoE between the United Kingdom and China

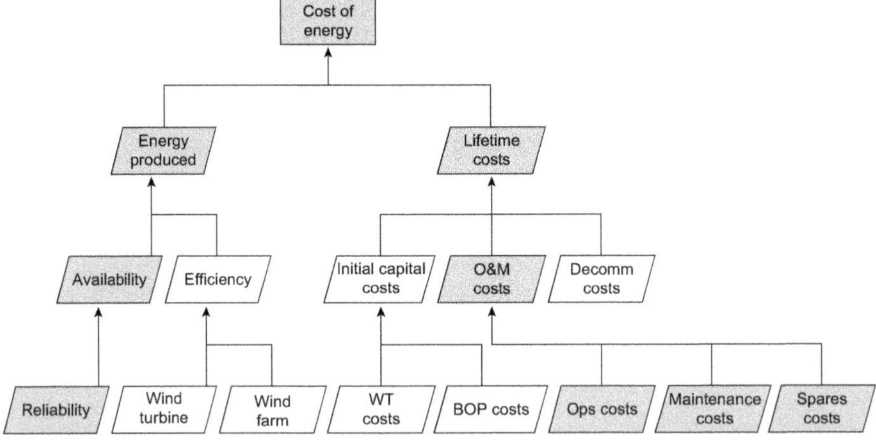

Figure 1.14   Structure of cost of energy, showing highlighted in grey areas of interest for this book [Source: [13]]

design and operations of the WT has been presented in Reference 13 and is shown in Figure 1.14, as the focus of this book is on the highlighted areas of the diagram.

1.6.4   O&M costs

The estimated cost of offshore wind energy varies depending on the site and project, but Section 1.6.2 shows that offshore wind projects are significantly more costly than onshore [4]. As WT designs become adapted to offshore

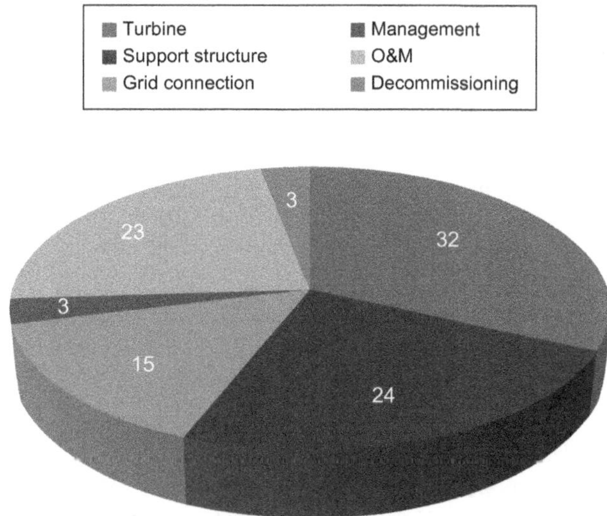

Figure 1.15   Typical cost breakdown for an offshore wind farm in shallow water

conditions, the achievement of a favourable economic solution depends upon controlling the wind farm system full life-cycle cost. Figure 1.15 illustrates a breakdown of typical total system costs for an offshore wind farm in shallow water [14]. Much of the price premium now being paid for offshore wind can be attributed to the WT Foundation, Grid Connection and Operation and Maintenance (O&M).

O&M for offshore wind farms is more complex than onshore. As a consequence, O&M percentage costs for some European offshore wind farms vary from 18% to 23%, much higher than the measured 12% for onshore projects [8]. Offshore conditions require more onerous erection and commissioning operations; meanwhile accessibility for offshore routine servicing and maintenance is a major issue. During winter, a whole wind farm may be inaccessible for many days due to harsh sea, wind or visibility conditions. Even given favourable weather, O&M tasks are more costly than onshore, being influenced by distance offshore, site exposure, wind farm size, WT reliability and maintenance strategy. Offshore conditions require special lifting equipment to install and change out major sub-assemblies, which may not be available at short notice or be locally sourced. Therefore, advanced techniques are needed to plan maintenance, using data from the Supervisory Control Data Acquisition (SCADA) and Condition Monitoring Systems (CMS) fitted to the WT, requiring a thorough knowledge of offshore conditions, qualitative physics theory and other design tools to predict failure modes in less conventional ways than has hitherto been done. Offshore remote monitoring and visual inspection become much more important to maintain appropriate WT availability and capacity factor levels.

1.6.5   Effect of reliability, availability and maintenance on cost of energy

Equation (1.11) for *CoE* can be expressed as a function of λ and μ allowing us to see the effect of reliability and maintenance on *A* and *CoE* as follows:

$$CoE = \frac{ICC \times FCR + O\&M(\lambda, 1/\mu)}{AEP(A(1/\lambda, \mu)} \tag{1.16}$$

Reductions in failure rate λ, will improve reliability *MTBF*, $1/\lambda$, and availability, *A*, therefore reducing O&M costs. Reductions in downtime *MTTR* will improve maintainability, μ, and availability, *A*, therefore also reducing O&M. As a consequence, *CoE* will also reduce as λ and μ improve.

1.6.6   Previous work

Professor J. Schmid published the first data on European WT reliability [1]. The EU FP7 ReliaWind project [15] prepared a report on the previous literature on WT reliability [16].

1.7   Roles

1.7.1   General

There are many stakeholders within the task of developing, building and operating offshore wind farms, whose actions define and shape our ability to achieve the objectives of that farm. Those objectives are to generate electricity reliably from the wind's renewable source at competitive prices and provide an acceptable return to each of the stakeholders. This book concerns the operation of the wind farm, once built, and the vital task of ensuring that the planned wind farm returns are extracted in an efficient and predictable way. The following describes the role of each of these major stakeholders so that the reader can understand their influence upon the planned process.

1.7.2   Regulator

In the United Kingdom, the regulator, the Office of the Gas and Electricity Markets (OFGEM), sets the market landscape for offshore wind. A particularly important aspect of this has been the development of the role for Offshore Transmission Operators (OFTO) ensuring that offshore wind farms will have a secure and flexible connection asset to transport the power into the onshore transmission grid. The long-term availability of the OFTO's connection asset and its reliability will be essential to the achievement of offshore wind farms objectives, but its technical reliability will be outside the scope of this book.

1.7.3   Investors

Investors in offshore wind include banks, energy companies and landowners, including the Crown Estate in the United Kingdom, which has licensed the offshore areas for development. In some ways the issues of reliability and availability of the wind farm asset are of most importance to the investors, since this is the means by which their investment can be reliably and predictably repaid with the required return. The difficulty for investors, in this emerging technology, is to understand the technical issues involved so that the right parameters can be defined for their investment. The object of this book is to explain the technical issues of offshore wind farm reliability and availability for them to be able to define their parameters more precisely.

1.7.4   Certifiers and insurers

Certifiers, such as Germanischer Lloyd and Det Norsk Veritas, are responsible for ensuring that WT designs and their associated marine structures are adequately certified to meet the IEC standards. Project insurers are also important participants as they determine the premium necessary to insure large offshore projects. An important aspect of these processes is imposing the necessary Health and Safety (H&S) regime on the installation and operational phases of the project to ensure that the human risks are acceptable.

These processes were developed for the onshore industry and have proved successful in ensuring that machines and structures are sound and safe investments. The processes are even more important offshore, where the environment is more challenging. However, this has meant that WT designs have focused on meeting safety and certification requirements more than production requirements.

1.7.5   Developers

Developers of offshore wind farms are emerging as consortia of investors, energy companies, WT manufacturers and operators. Their objective is to gain a return on the development of wind farm generation assets that are subsequently sold onto long-term operators such as the main electricity generating companies. Because of the scale and complexity of the offshore asset these consortia are drawing in long-term investors as part of the development team and that requires financial experts to have a better understanding of the technical issues concerned.

A major part of the deployment of offshore wind farms depends upon the marine installation assets, including port and docking facilities, installation vessels, maintenance vessels fleet and the manpower and infrastructure to manage and operate these assets, which are usually provided by civil and marine engineering businesses, who are starting to become important members of wind farm developer consortia.

1.7.6   Original equipment manufacturers

The principal OEMs involved in the wind farm are the WT OEMs. But the wind farm is a complex generation, collection and transmission asset with a substantial Balance of Plant (BOP), which is drawing in cable and transmission OEMs as well.

The actions of the regulator are tending to push the transmission OEMs to participate in the OFTO activity, but they still have a significant financial, management and technical role in the collection and offshore substations of the offshore wind farm.

1.7.7   Operators and asset managers

The operators of offshore wind farms are large energy companies providing electricity into the transmission grid.

Most of these operators are broad-technology generators with fossil- and nuclear-fired and renewable generation assets. In view of the technical complexity of offshore wind assets a few specialised offshore operators are developing, particularly from the Scandinavian market, and are developing their expertise to match their existing assets in onshore wind, hydro and gas-fired generation.

It seems likely in the future that more specialised operators will develop but the size and complexity of offshore wind assets means that these will be large operators with a large international portfolio of assets, which will be developed to balance their exposure and risk in the offshore wind sector.

As the industry matures, the current certification- and safety-oriented approach is likely to change, as the more stringent demands for return on the larger capital outlays for capital projects encourages a more vigorous production-oriented approach. In this stage of development of the industry, the interaction between operators, asset managers, certifiers, insurers and investors will be strengthened.

1.7.8   Maintainers

Maintainers work for a variety of the wind farm stakeholders. Offshore WT OEMs have large, experienced service departments of maintainers, with knowledge of the O&M of their WTs onshore and offshore. They have access to the SCADA data streaming from wind farms with their machines during the commissioning and warranty periods. Some WT OEMS have data centres where all their WTs data can be viewed by service and design staff. They also have detailed knowledge of the development of their own WTs through prototype tests, supply chain development and production tests. Their staffs are trained on their machines and have built up a detailed personal knowledge of the idiosyncrasies of individual WT types. This expertise is deployed during the warranty period, regulated by the project contract. WT OEMs have some knowledge of the long-term life of the wind generation asset but generally lack asset management experience. For some WT OEMs, this may change with time as they recognise the benefit to their business of the O&M market and the importance to the developers and operators of through-life performance.

Operators also have substantial experience of wind farm operation, different in nature with that of the WT OEM, being more focused on production needs and the through-life performance of the asset. They will have their own management and some of their own O&M staff but may rely upon sub-contractors and the WT OEM for some of that support. However, they frequently lack detailed knowledge of individual wind farm equipment and rely, in large part, upon the warranty period to gain that knowledge and experience.

Operators may opt to continue with a maintenance contract with the WT OEM after the completion of the warranty period. But as offshore wind farm operators are large, with experience of many wind farms, many will opt to undertake their own O&M under their own direction to impose their own asset management objectives upon the wind farm and ensure long life.

Wind farm maintenance relies heavily upon the skill of the management and staff carrying out this highly skilled activity. Wind farm design, choice of WT, availability of appropriate access assets, spares and tools can facilitate the activity but success is impossible without staffs who are well trained in H&S and the technology of the asset. This is an important issue that will be addressed later.

1.8   Summary

The development of large onshore wind farms has been accelerating around the world over the last 20 years so that wind farms >100 MW in rating are now commonplace and the world's installed capacity is >238 GW with an annual energy production of >345 TWh. Confidence with large onshore wind operations has encouraged nations and developers to start developing larger offshore wind farms over the last 10 years.

The lead is currently being taken in Europe, in the North, Baltic and Irish Seas, with the United Kingdom currently having the largest installed offshore capacity with a potential annual energy production of >800 GWh and the largest offshore wind farm rated at 300 MW from 100 WTs.

China has also made a large commitment to offshore wind having installed 133 MW of OWTs, and it seems that, with its large south-eastern coastal electricity load, well-developed grid in those areas and good offshore wind resource, we are likely to see a large expansion in the near future.

The United States has started to consider the opportunities on its eastern seaboard and this could also be a region of high growth.

Economic analyses of European offshore wind sites to date have shown that the WT installation cost is approximately 100% more than onshore, the *CoE* is about 33% more than onshore, whilst the O&M cost is 18–23% more than onshore, all depending upon the offshore wind location, changing as lessons are learnt in the field.

There are a number of roles in the offshore wind industry and these have been clearly set out in this chapter.

1.9   References

[1]   Schmid J., Klein H.P. *Performance of European Wind Turbines*. London and New York: Elsevier, Applied Science; 1991. ISBN 1-85166-737-7

[2]   Tavner P.J., Xiang J.P., Spinato, F. 'Reliability analysis for wind turbines'. *Wind Energy*. 2006;**10**(1):1–18

[3]   Ribrant J., Bertling L. 'Survey of failures in wind power systems with focus on Swedish wind power plants during 1997–2005'. *IEEE Transactions On Energy Conversion*. 2007;**22**(1):167–73

[4]   van Bussel G.J.W., Schöntag C. 'Operation and maintenance aspects of large offshore windfarms'. *Proceedings of European Wind Energy Conference, EWEC1997*. Brussels, Belgium: European Wind Energy Association; 1997.

[5]   Beurksens J. (ed.). *Converting Offshore Wind into Electricity-the Netherlands Contribution to Offshore Wind Energy Knowledge*. Delft, Netherlands: Eburon Academic Publishers; 2011. ISBN 978-90-5972-583-6

[6]   UK DTI (Department of Trade and Industry) and BERR (Department for Business Enterprise and Regulatory Reform) (2007 and 2009) Offshore wind capital grants scheme annual reports. London, UK: DECC & BERR

[7]   Faulstich S., Lyding P., Tavner P.J. 'Effects of wind speed on wind turbine availability'. *Proceedings of European Wind Energy Conference, EWEA2011*, Brussels: European Wind Energy Association; 2011

[8]   van Bussel G.J.W., Henderson A.R., Morgan C.A., Smith B., Barthelmie R., Argyriadis K., *et al. State of the Art and Technology Trends for Offshore Wind Energy: Operation and Maintenance Issues* [Online], Delft University of Technology. European Commission; 2001

[9]   EWEA. *The Facts*. EWEA; 2009. ISBN 978-1-84407-710-6

[10]   OECD. *Projected Costs of Generating Electricity*. IEA, OECD, NEA Report, Paris, France; 2005

[11]   Walford C.A. *Wind Turbine Reliability Understanding and Minimizing Wind Turbine Operation and Maintenance Costs*. Sandia Labs, Albuquerque, USA, Report SAND, 2006

[12]   Feng Y., Tavner P.J., Long H. 'Early experiences with UK round 1 offshore wind farms'. *Proceedings of Institution of Civil Engineers, Energy*. 2010;**163**(EN4):167–81

[13]   Jamieson P. *Innovation in Wind Turbine Design*. Chichester, UK: John Wiley & Sons Ltd; 2011. ISBN 978-0-470-69881-2

[14]   Byrne B.W., Houlsby G.T. 'Foundations for offshore wind turbines'. *Philosophical Transactions of the Royal Society London*. 2003;**361**(A):2909–30

[15]   ReliaWind. Available from http://www.reliawind.eu/ [Last accessed 8 February 2010]

[16]   Arabian-Hoseynabadi H. ReliaWind Project Deliverable. D.1.1 Literature Review, November 2008. Available from http://www.reliawind.eu/files/publications/pdf_20.pdf [Accessed 10 May 2012]

Chapter 2

Reliability theory relevant to offshore wind turbines

2.1 Introduction

A modern, 2 MW WT is a large steel and concrete structure on which is mounted a complex electro-mechanical generating machine. The reliability of the whole device is dependent on epistemic uncertainty affecting

- the structural reliability, for which predicted failure rates are $<10^{-4}$ failures/year, and the probabilistic spread of those low failure rate events needs to be considered;
- the electro-mechanical reliability, which is subject to the normal vagaries of rotating machinery and can be predicted using measured constant failure rates for individual sub-assemblies ranging from 10^{0} to 10^{-3} failures/year;
- the control system reliability, which depends on the environment, electro-mechanical issues and the reliability of the software contained within the control system.

Such analysis is made more complex because the turbine is also subject to aleatory uncertainty due to the stochastic effects of the weather itself, the wind from which the machine extracts energy and, in the case of OWTs, the combined effects of wind and waves on the structure and of corrosion.

In order to understand and predict these effects there must be a detailed understanding of reliability theory, a relevant textbook on the subject is Reference 1.

To track changes of reliability with time during the different operational phases of a product, reliability growth models have been developed most notably using the Crow-AMSAA (Army Materiel Systems Analysis Activity) model [2]. The same model can be applied on failure data collected from the field to investigate whether product reliability stays constant or shows an improvement or deterioration with time.

2.2 Basic definitions

The reliability of a sub-assembly is defined as the probability that it will meet its required function under stated conditions for a specified period of time. This definition of reliability breaks down into four essential elements:

- Probability

- Required function
- Time variable
- Operational conditions for adequate performance

The complement of reliability, unreliability, is related to a failure intensity function, $\lambda(t)$, to be defined later.

This reliability definition experiences difficulties as a measure for continuously operated systems, such as WTs, which tolerate failures that can be repaired. Then a more appropriate measure is availability, defined as the probability of finding the system in the operating state at some time into the future. This definition then reduces to only two elements:

- Operability
- Time

Failure is the inability of a sub-assembly to perform its required function under defined conditions; the item is then in a failed state, in contrast to an operational or working state.

A non-repairable system is one that is discarded after a failure. Examples of non-repairable systems are small batteries or light bulbs.

A repairable system is one that, when a failure occurs, can be restored into operational condition after any action of repair, other than replacement of the entire system. Examples of repairable systems are WTs, car engines, electrical generators and computers.

Repair actions can be an addition of a new part, exchange of parts, removal of a damaged part, changes or adjustment to settings, software update, lubrication or cleaning.

2.3   Random and continuous variables

The random variable, in the context of WT reliability, is failures X recorded discretely against a continuous variable, such as time. Is it always appropriate to use calendar time as the continuous variable? Calendar time may be convenient but is not necessarily the best for reliability analysis, for example

- time on test seems more appropriate;
- turbine rotations may also be more appropriate, especially for the aerodynamic and transmission sub-assembly reliability;
- energy generated by the WT, GWh, may also be more appropriate, especially for electrical sub-assemblies reliability.

Operators usually cannot measure time on test because they cannot easily keep track of the date of origin of the WT but they can easily measure the number of failures in an interval of time, which is called censored data.

An example of such differences in the random variable is shown in Figure 2.1 [3] where identical failure data from the large German WSD (Wind-stats database for Germany) survey, referred to in Chapter 3, are in this case

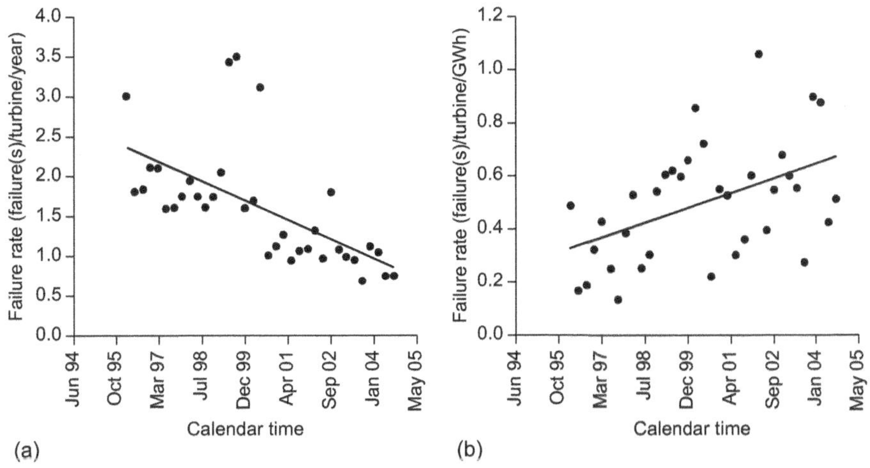

*Figure 2.1   Comparison of plots of identical failure data from the WSD survey
plotted as failure(s)/turbine/year or as failure(s)/turbine/GWh.
(a) Failure(s)/turbine/year vs time; (b) failure(s)/turbine/GWh vs time
[Source: [3]]*

plotted against calendar time in terms of either failure(s)/turbine/year or failure(s)/turbine/GWh. The former, in Figure 2.1(a), shows improving failure rate with time, whereas the latter in Figure 2.1(b) shows a wider variance but an increasing number of failures per GWh generated. The latter probably shows the extent of small but significant failures occurring in the growing number of larger, more technically complex WTs.

What this shows us is that the method of collecting and presenting data is important. The choice of continuous variable against which the random variable, X, is to be collected is important:

- The random variable X can be presented in different ways.
- The discrete or continuous variable can be presented in different ways.
- Whether it is to be calendar time, time on test, GWh or rotations needs to be selected based on the interpretation to be made.
- Plotting X in different ways against different discrete or continuous variables reveals different information.
- Whether the component on which the data are being collected is repairable or non-repairable needs to be determined.
- If the data collection method is good and the variable chosen appropriately, then the statistical data of the random variable X collected should yield robust reliability information.
- If not, the reliability information may be faulty.

Now we can consider probability distributions of a random variable.

2.4   Reliability theory

2.4.1   Reliability functions

The following equations and mathematical relationships between the various reliability functions do not assume any specific failure distribution and are equally applicable to all probability distributions used in reliability evaluation. Consider N_0 identical components are tested:

$$N_s(t) = \text{number surviving at time } t \tag{2.1}$$

$$N_f(t) = \text{number failed at time } t \tag{2.2}$$

Therefore,

$$N_s(t) + N_f(t) = N_0 \tag{2.3}$$

At any time, t, the survivor or reliability function, $R(t)$, is given by

$$R(t) = \frac{N_s(t)}{N_0} \tag{2.4}$$

Similarly, the probability of failure or cumulative distribution function or unreliability function, $Q(t)$, is given by

$$Q(t) = \frac{N_f(t)}{N_0} \tag{2.5}$$

where

$$R(t) = 1 - Q(t) \tag{2.6}$$

The failure density function, $f(t)$, is given by

$$f(t) = \frac{1}{N_0}\left(\frac{dN_f(t)}{dt}\right) \tag{2.7}$$

Failure intensity or hazard rate function:

$$\lambda(t) = \frac{1}{N_s(t)}\left(\frac{dN_f(t)}{dt}\right) \tag{2.8}$$

$$\lambda(t) = \frac{1}{R(t)}\left(\frac{dR(t)}{dt}\right) \tag{2.9}$$

Failure density function is normalised to the number of survivors, $\lambda(t)$ (see Figure 2.2).

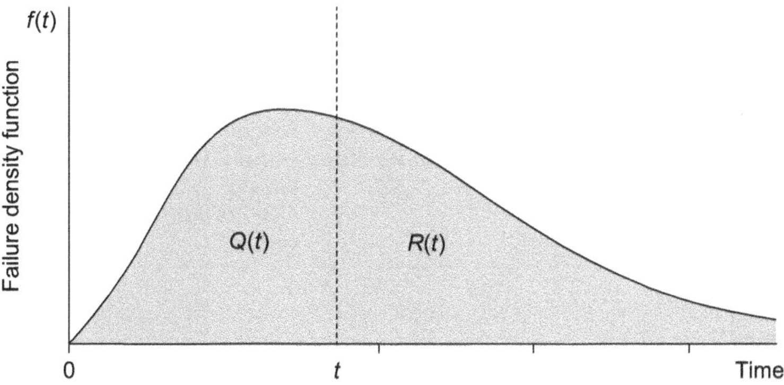

Figure 2.2   Failure density function against time showing reliability R(t) and Q(t)

The special case in which λ is constant and independent of time is an exponential distribution, and the hazard rate becomes the failure rate. Where the hazard rate/failure rate $\lambda(t)$ = (number of failure per unit time/number of components exposed to failure):

$$R(t) = 1 - Q(t) \tag{2.10}$$

$$f(t) = \frac{dQ(t)}{dt} = \frac{-dR(t)}{dt} \tag{2.11}$$

or

$$Q(t) = \int_0^t f(t)dt \tag{2.12}$$

and

$$R(t) = 1 - \int_0^t f(t)dt \tag{2.13}$$

The total area under the failure density function must be unity. Therefore,

$$R(t) = \int_0^\infty f(t)dt = 1 \tag{2.14}$$

2.4.2   Reliability functions example

The following is an example of these methods using contrived data from a large offshore wind farm and the example is based upon one given in Reference 4.

The example considers a large offshore wind farm of 1000 WTs and for the sake of this example they are each non-repairable. There is a steady failure of these WTs. Table 2.1 records the cumulative failures and survivors over a period of

Table 2.1   Record of failures of 1000 non-repairable WTs in an offshore wind farm

Time interval (years)	Number of failures in each interval	Cumulative failures, N_f	Number of survivors, N_s	Failure density function, $f(t)$	Unreliability function or cumulative failure distribution, $Q(t)$	Reliability or survivor function, $R(t)$	Failure intensity or hazard rate, $l(t)$
0	140	0	1000	0.140	0	1.000	0.151
1	85	140	860	0.085	0.140	0.860	0.104
2	75	225	775	0.075	0.225	0.775	0.102
3	68	300	700	0.068	0.300	0.700	0.102
4	60	368	632	0.060	0.368	0.632	0.100
5	53	428	572	0.053	0.428	0.572	0.097
6	48	481	519	0.048	0.481	0.519	0.097
7	43	529	471	0.043	0.529	0.471	0.096
8	38	572	428	0.038	0.572	0.428	0.093
9	34	610	390	0.034	0.610	0.390	0.091
10	31	644	356	0.031	0.644	0.356	0.091
11	28	675	325	0.028	0.675	0.325	0.090
12	40	703	297	0.040	0.703	0.297	0.144
13	60	743	257	0.060	0.743	0.257	0.264
14	75	803	197	0.075	0.803	0.197	0.470
15	60	878	122	0.060	0.878	0.122	0.652
16	42	938	62	0.042	0.938	0.062	1.024
17	15	980	20	0.015	0.980	0.020	1.200
18	5	995	5	0.005	0.995	0.005	2.000
19	0	1000	0	0			
TOTAL 1000				**1**			

19 years calculating the failure density function, which sums to 1, and the hazard rate. So Table 2.1 records the reliability of this wind farm, while Figure 2.3 plots all these functions so that their nature can clearly be seen.

Figures 2.3(c) and (d) are the most interesting as they show respectively the failure density function, the area under which accumulates to 1, compare with Figure 2.2, and the hazard rate. This clearly shows the bathtub form given in Figure 2.4, with the early failures phase I, steady failure rate phase II and wear-out phase III. Particularly interesting is phase II where Figure 2.3(c) shows the failure density function decreasing exponentially, representing the random nature of failures in that phase. When the failure density function is normalised into the hazard rate in Figure 2.3(d) during phase II those random failures become a constant hazard or failure rate.

2.4.3   Reliability analysis assuming constant failure rate

The unreliability of repairable systems can be modelled in terms of failure intensity by the bathtub curve [5], which represents the three different phases of a population life, as shown in Figure 2.4.

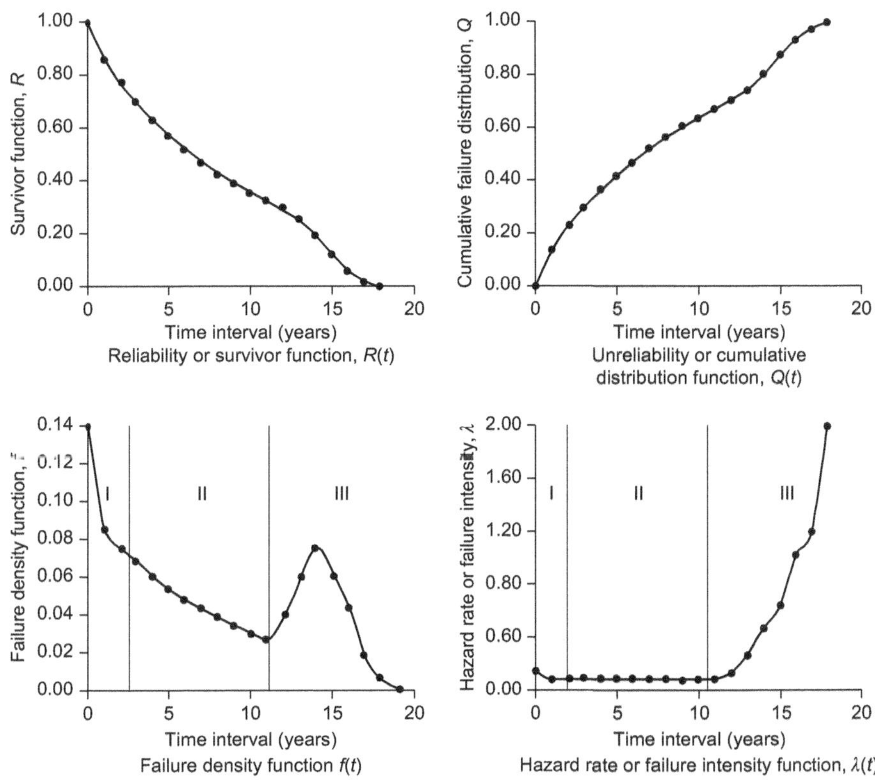

Figure 2.3   Reliability functions from a wind farm of 1000 non-repairable WTs

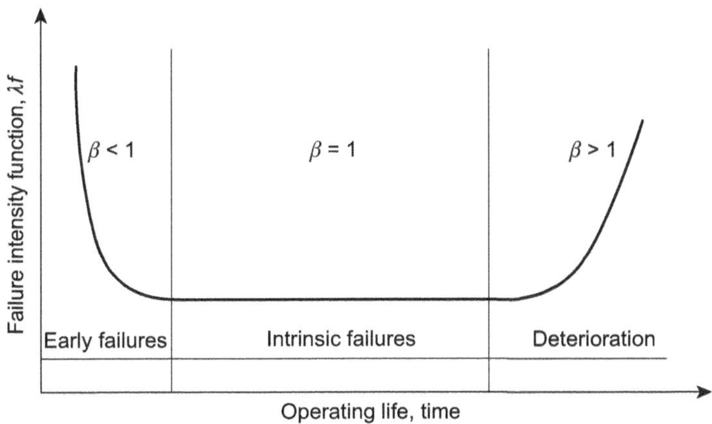

*Figure 2.4   The 'bathtub curve' for the intensity function showing how the
reliability varies throughout the life of repairable machinery*

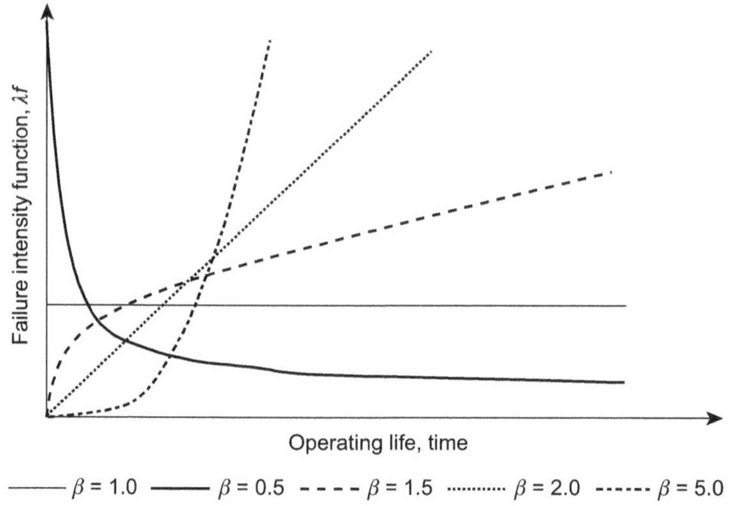

Figure 2.5   The power law function showing how the failure intensity varies with the shape parameter β

In turn, each phase of the bathtub curve can be modelled by a failure intensity function as shown in Figure 2.5.

This section is based on the concept of a bathtub curve (Figure 2.4) for a repairable system and its mathematical formulation, the power law process (PLP). The PLP is a special case of a Poisson process with a failure intensity function

$$\lambda(\tau) = \rho\beta t^{\beta-1} \tag{2.15}$$

β determines the trend of the curve, is dimensionless and is called the shape parameter, failure intensity changes with the shape parameter β.

ρ is a scale parameter, which has the unit year^{-1}. $\lambda(t)$ has units in this section of failures per item per year or year^{-1}, where an item can be a WT or a sub-assembly.

For $\beta < 1$ or $\beta > 1$, the curve shows, respectively, a downward or upward trend. When $\beta = 1$, the intensity function of the PLP is equal to ρ, the process represents the bottom of the bathtub curve, called the intrinsic failures phase, and λ is described as the average failure rate.

Elements of the reliability theory used to analyse the failure data are summarised in References 1 and 4–6 and in the next section.

2.4.4   Point processes

A point process is a stochastic model describing the occurrence of discrete events in time or space. In reliability analysis, failures of repairable systems can be described with point processes in the calendar time domain, for example hourly, quarterly or annually, or using an operational variable, like kilometres driven or number of flying hours.

A random variable $N(t)$ that represents for example the number of failure events in the interval $[0, t]$ is called the counting random variable. Subsequently, the number of events in the interval $(a, b]$ will be

$$N(a,b] = N(b) - N(a) \tag{2.16}$$

The point process mean function $\Lambda(t)$ is the expected number of failures, E, in the interval throughout time t:

$$\Lambda(t) = E[N(t)] \tag{2.17}$$

The rate of occurrences of failure $\mu(t)$ is the rate of change of expected number of failures

$$\mu(t) = \frac{d\Lambda(t)}{dt} \tag{2.18}$$

The intensity function $\lambda(t)$ is the limit of probability, P, of having one or more failures in a small interval divided by the length of the interval:

$$\lambda(t) = \lim_{\Delta t \to 0} P(N(t, t + \Delta t]) \geq \frac{1}{\Delta t} \tag{2.19}$$

If the probability of simultaneous failures is zero, which occur only where the mean function $\Lambda(t)$ is not discontinuous, then

$$\lambda(t) = \mu(t) \tag{2.20}$$

2.4.5 *Non-homogeneous Poisson process*

Assuming minimal repair, that is failed sub-assemblies are brought back to the same condition as just before the failure, the non-homogeneous Poisson process (NHPP) can be used to describe changes in reliability of repairable systems [5]. A counting process $N(t)$, that is the cumulative number of failures after operational or calendar time t, is a Poisson process if

$$N(0) = 0 \tag{2.21}$$

For any a < b ≤ c < d, the random variables $N(a,b]$ and $N(c,d]$ are independent. This is known as the independent increment property.

There is an intensity function λ such that

$$\lambda(t) = \lim_{\Delta t \to 0} \frac{(P(N(t, t + \Delta t]) = 1)}{\Delta t} \tag{2.22}$$

Note that if λ is constant then the process is homogeneous Poisson process (HPP).

Simultaneous failures are not possible

$$\lim_{\Delta t \to 0} \frac{(P(N(t, t + \Delta t]) \geq 2)}{\Delta t = 0} \tag{2.23}$$

The main property of NHPP is that the number of failures $N(a,b]$ in the interval $(a,b]$ is a random variable having a Poisson distribution with mean

$$\Lambda(a, b] = E[N(a, b]] = a\lambda(t)dt \tag{2.24}$$

2.4.6 Power law process

An NHPP is called a PLP if the cumulative number of failures through time t, $N(t)$ is given by

$$N(t) = \rho t^{\beta} \tag{2.25}$$

Therefore, the expected number of failures for a specific time interval $[t_1, t_2]$ will be

$$N[t_1, t_2] = N(t_2) - N(t_1) = \rho(t_2^{\beta} - t_1^{\beta}) \tag{2.26}$$

The intensity function is then

$$\lambda(t) = \frac{dN(t)}{dt} = \rho(t_2^{\beta} - t_1^{\beta}) \tag{2.27}$$

One of the advantages of using the PLP model for repairable systems is that its intensity function (2.12) is flexible enough to represent separately the three different phases of the bathtub curve (see Figure 2.4), based on the value of the shape parameter β, as described in Table 2.2.

Table 2.2   Values of β for different failure intensities

Value of β	Failure intensity	Reason	Model type
$\beta < 1$	Decreasing with time design	Improvements/Alterations on field	NHPP
$\beta = 1$	Constant with time $\lambda(t) = \rho$	No major design modifications – wear and tear not apparent yet	HPP
$\beta > 1$	Increasing with time normal	Deterioration of materials/ accumulated stresses	NHPP

2.4.7 Total time on test

The variable t that appears in the various equations of the Crow-AMSAA model [2] represents the time to a point process but it differs from calendar time, as reported in the failures tables of WSDK (Windstats database for Denmark), WSD and LWK (Landwirtschaftskammer Schleswig-Holstein database for Germany). Reliability growth, as well as other reliability analysis, is normally carried out on the basis of specific tests made on sub-assemblies under investigation. For a repairable system, the test is stopped after a failure or a planned inspection and the number of running hours elapsed since the previous failures are recorded. After a number of failures have been accumulated, failure data are interpolated with a mathematical model, like the Crow-AMSAA, to verify the achieved reliability, or, using the terminology of the military standard, the 'demonstrated reliability'. The independent variable *t* of the plot is the cumulative quantity called the total time on test (*TTT*), which is the integral of the number of running hours of the entire population for the observed period. In this way the hours of inactivity are not included in the evaluation of the *TTT*. Using *TTT* rather than calendar time presents advantages and disadvantages, and the meaning of *TTT*, for WT failure data, must be clarified [7]. First, it is in the nature of reliability engineering to deal with running hours rather than calendar time. This distinguishes a reliability analysis from an availability analysis. In this case the age of many electro-mechanical systems can be measured with the number of cycles completed or the total running hours and often this differs substantially from the calendar age. Nevertheless, the calendar time plays an important role in reliability studies where chemical–physical properties deteriorate with time, for example the insulating property of a dielectric. For data sets like LWK, WSD or WSDK, the *TTT* in a certain interval i, ΔTTT_i, is calculated by multiplying the number of WTs, N_i, by the number of hours in the interval, h_i. The recorded total hours lost from WT production, l_i, in that interval are then subtracted, when this information is available. In these surveys, this data included only out of service time, rather than time when the WT was unable to operate for lack of wind. The aggregated *TTT* up to an arbitrary time cell k, t_k, is then

$$t_k = \Sigma_{i=1}\Delta TTT_i = \Sigma_{i=1}N_i(h_i - l_i) \tag{2.28}$$

To calculate the *TTT* for the LWK, WSD or WSDK data, three considerations are necessary.

For each time interval, the WTs in the survey are considered representative of the entire of population. Therefore, the sample reliability for each time interval is assumed to represent the reliability of the entire population. This hypothesis is necessary to overcome one of the major deficiencies of the data, the variable number of WTs in each time interval. In reality, any reliability improvement or deterioration spreads throughout the population with a certain rate, indicated by the shape parameter b, as long as sample WTs are assumed randomly chosen from the entire population and the usage of each WT in the population is similar.

Using *TTT* has the effect of stretching the curve on the abscissa. Since *TTT* depends on the number of turbines considered, it has no absolute meaning, as calendar time would have. The abscissa *t* has significance only for the WT population being examined; however, by showing the cursor at the right of Figure 2.6, calendar time can be inferred.

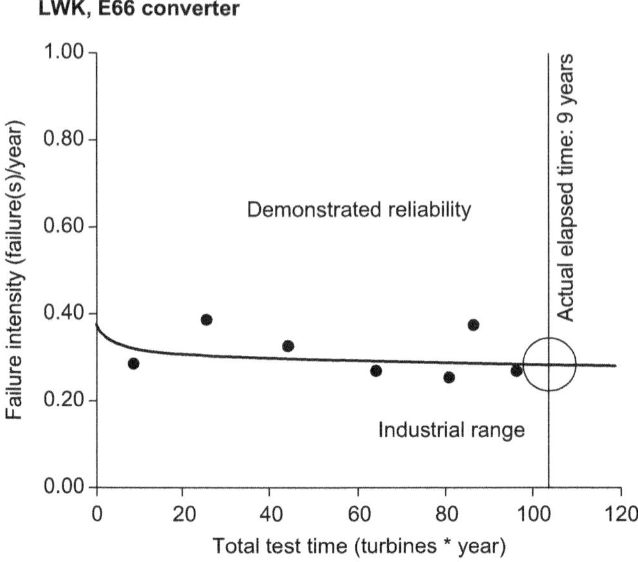

Figure 2.6   Presentation of failure intensity using total time on test, TTT, showing demonstrated reliability for a sub-assembly with early failures [Source: [7]]

As the intensity function interpolates data on *TTT* rather than calendar time, the fit produced is intrinsically weighted by the number of turbines in each period. A larger number of WTs results in a larger *TTT* interval and the fit constraint is stronger. When *TTT* is used rather than calendar time, the abscissa stretches to a longer interval for more WTs surveyed and the scale parameter increases. In cases of early or constant failures the most important result is the demonstrated reliability, as shown in Figure 2.6.

2.5   Reliability block diagrams

2.5.1   General

Individual sub-assemblies can be represented in the process of reliability modelling and prediction (RMP), using the methods above, by reliability block diagrams (RBD) in a set and then connected in series or parallel to represent their functionality. Figure 2.7 shows possible arrangements for two reliability blocks.

2.5.2   Series systems

Sub-assemblies in a set are said to be in series, from a reliability point of view, if they must all work for system success and only one needs to fail for system failure.

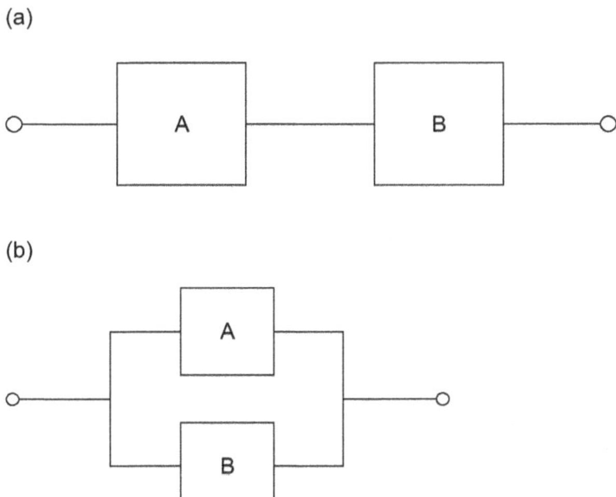

*Figure 2.7   Representation of sub-assemblies in a reliability block diagram.
(a) Series components; (b) parallel components*

Consider a system consisting of two independent components A and B connected in series, for example a gear train.

$$R_p = \Pi R_i \tag{2.29}$$

This equation is referred to as the product rule of reliability.

Let R_a and R_b be the probability of successful operation of the individual sub-assemblies A and B, respectively, in Figure 2.7(a), and R_s be the probability of successful operation of the series set.

Let Q_a and Q_b be the probability of failure of sub-assemblies A and B, respectively:

$$R_s = R_a \times R_b \tag{2.30}$$

Example: A gearbox consists of six successive identical gear wheels, all of which must work for system success. What is the system reliability of the series set if each gearwheel has a reliability of 0.95? From the product rule

$$R_s = 0.95^6 = 0.7350$$

2.5.3   *Parallel systems*

Sub-assemblies in a set are said to be in parallel, from reliability point of view, if only one needs to be working for system success or all must fail for system failure.

Consider a system consisting of two independent components A and B, connected in parallel (Figure 2.7(b)), for example two lubrication oil pumps for a gearbox connected in parallel. From a reliability point of view, the requirement is that only one sub-assembly has to be working for system success.

Again let R_a and R_b be the probability of successful operation of individual sub-assemblies and R_p be the probability of successful operation of the parallel set. Let Q_a and Q_b be the probability of failure of sub-assemblies A and B, respectively:

$$Q_p = \Pi Q_i \tag{2.31}$$

$$R_p = 1 - \Pi Q_i \tag{2.32}$$

Example: A system consists of four pumps in parallel each having reliabilities of 0.99. What is the reliability and unreliability of the parallel set?

$$Q_p = (1 - 0.99)^4 = 0.01^4 = 0.00000001$$
$$R_p = 1 - Q_p = 0.99999999$$

2.6 Summary

This chapter has presented the essential reliability mathematics necessary to understand the data collected from WTs and wind farms and presented in this book. It shows that simple methods can be used to extract essential information and the overall results that can be obtained.

However, care must be taken in manipulating the data to ensure that interpretations are sound.

2.7 References

[1] Birolini A. *Reliability Engineering, Theory & Practice*. New York: Springer; 2007. ISBN 978-3-538-49388-4

[2] *MIL-HDBK-189: Reliability Growth Management*. Washington: US Department of Defense; 1981

[3] *Windstats (WSD & WSDK) quarterly newsletter, part of WindPower Weekly, Denmark*. Available from http://www.windstats.com [Last accessed 8 February 2010]

[4] Billinton R., Allan R.N. *Reliability Evaluation of Engineering Systems: Concepts and Techniques*. 2nd edn. New York & London: Plenum Press; 1992. ISBN-13: 978-0306440632

[5] Rigdon S.E., Basu A.P. *Statistical Methods for the Reliability of Repairable Systems*. New York: John Wiley & Sons; 2000

[6] Goldberg H. *Extending the Limits of Reliability Theory*. New York: John Wiley & Sons; 1981

[7] Spinato F. *The Reliability of Wind Turbines*. Doctoral thesis, Durham University; 2008

Chapter 3
Practical wind turbine reliability

3.1 Introduction

This chapter describes the reliability of current WTs using research from References 1–3, based on onshore WTs, with some additional information from OWTs from Reference 4. Figure 1.2 showed the gross failure rate trend for onshore turbines, and at this stage it is important to define what a failure can be.

WTs are unmanned robotic devices and it is relatively rare that their stoppages can readily be classified as a failure, with the possible exception of a major gearbox, generator or blade failure, where the cause of failure is obvious. More normally, the WT is stopped because its controller has detected an operational condition outside the WT's safe envelope. This is usually the result of an unacceptable operational condition, such as an over-temperature, over-speed or pitch problem, and the control system disconnects the WT from the grid, puts it into the emergency feather condition (EFC) and the turbine comes to a stop. The fault can be resolved by either:

- an automatic restart; or
- a manually initiated remote restart; or
- a site visit by a WT technician, who may merely initiate a local restart; or
- a site visit by a WT technician triggering a repair operation, which then allows the WT to be restarted.

In each case these cause a stoppage, and the figures shown in Figure 1.2 can really be regarded as stoppage rates rather than failure rates. The surveys referred to in References 1–3 are concerned with stoppages >24 hours. Therefore, they constitute serious stoppages, which usually cannot be resolved by an automatic, remote or local restart, with a downtime of at least 24 hours. They usually, therefore, involve some form of damage, the exact nature of which cannot be identified by the WT OEM or operator until after a faulty sub-assembly has been replaced or repair work done.

Therefore, to determine a WT's reliability we must have a working knowledge of the measured stoppage or failure rate, λ, which allows us to determine an $MTBF = 1/\lambda$. To understand availability we need to know the stoppage or downtime, which makes up the logistic delay time, LDT, and $MTTR = 1/\mu$, from the repair rate, μ, allowing us to determine the availability, $A = MTBF/(MTBF + MTTR + LDT)$, see (1.3)–(1.10).

Knowledge of WT failure rates allows us to compare WT reliability performance and calibrate the contribution made to their unreliability of particular sub-assemblies. In this way the future performance of WTs can be improved by maintenance.

Interestingly, if a survey shows a low failure rate with long *MTTR* or stoppage time, this may result in the same WT availability as much higher failure rates with lower *MTTR*. For example, a survey showing 97% availability WTs with a failure rate of 1 failure(s)/turbine/year for $\geq$24 hours stoppages will show the same availability as a survey of WTs with a failure rate of 24 failure(s)/turbine/year for $\geq$1 hour stoppages.

3.2   Typical wind turbine structure showing main assemblies

The basic structure of a modern three-blade, upwind HAWT is exemplified by Figure 3.1.

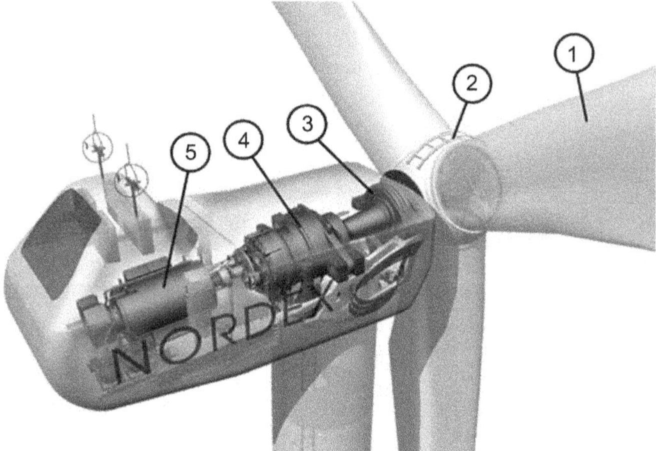

Figure 3.1   Structure of a modern three-bladed, upwind HAWT. (1) Blades; (2) hub containing pitch mechanism; (3) main bearing; (4) gearbox; (5) generator [Source: Nordex]

The main assemblies are shown, but there is a large variety of modern designs and it is important to capture failures, their precise locations in the structure and record their effect on a WT's reliability availability. For operational purposes the WTs are fitted with SCADA and CMS systems that automatically collect data from transducers and alarm circuits distributed around the WT structure and enable the WT to operate automatically within its operating envelope.

3.3   Reliability data collection

The wind industry has not yet standardised its methods of reliability data collection, whereas the oil and gas industry has done so [5]. However, an early wind industry reliability study [6], Wissenschaftlichen Mess- und Evaluierungsprogramm database (WMEP) in Germany, developed a prototype data collection system described in the reference. In particular, each incident occurring on a WT was described by a standardised Operators Report form, which is given in Appendix 3.

The EU FP7 ReliaWind Consortium [7] developed a standard approach to data collection based on WMEP and other work, described in Appendix 2, catering specifically for larger wind farms and making use of both automatic but filtered SCADA data and maintainers' logs, rather than Operators Reports. Within a data collection system, it is necessary to define the structure or taxonomy of the plant from which data are to be collected, for the wind farm and the individual WTs. The taxonomy will define the detail of the data to be collected, the more detailed that taxonomy the more detailed will be the data collected. WTs are fitted with SCADA systems that collect data from around the WT structure, as described in the previous section. This structure should coincide with the planned taxonomy for collecting reliability data as the SCADA system is already collecting such data automatically, albeit in greater volume per unit time than that collected by Operators Reports. The WT taxonomy is described in the next section.

3.4   Wind turbine taxonomies

The taxonomy of a WT is the standardised structure needed so that we can define accurately failure locations and identify where we are to concentrate maintenance and repair activity to maximise availability. In Reference 9, a power industry standard has been applied to a WT to derive a taxonomy and naming of parts for the wind industry. The ReliaWind Consortium also developed a standardised taxonomy that reflects standards and caters specifically for large wind farms. That taxonomy, see Section 11.2.3, is based upon a five-level system as follows:

- System, which could be the wind farm including WTs, substation and cables
- Sub-system, which could be an individual WT in that wind farm
- Assembly, which could be, for example, the gearbox in that WT
- Sub-assembly, which could be, for example, the high-speed shaft in that gearbox
- Component, which could be the high-speed bearing on that shaft

The document also prescribes the way in which reliability data should be collected from wind farms based on the approach of Reference 10. This taxonomy will be used throughout the rest of this book.

3.5   Failure location, failure mode, root cause and failure mechanism

The WT taxonomy will allow us to identify accurately in a reliability survey a failure location, but from a reliability point of view we need also to understand the root cause of failure and the failure mechanism that links the two. Figure 3.2 shows the relationship between the root cause and the failure mode, while Figure 3.3 helps make this clear showing an example of the linkage between the root cause and the failure mode of a WT main shaft failure.

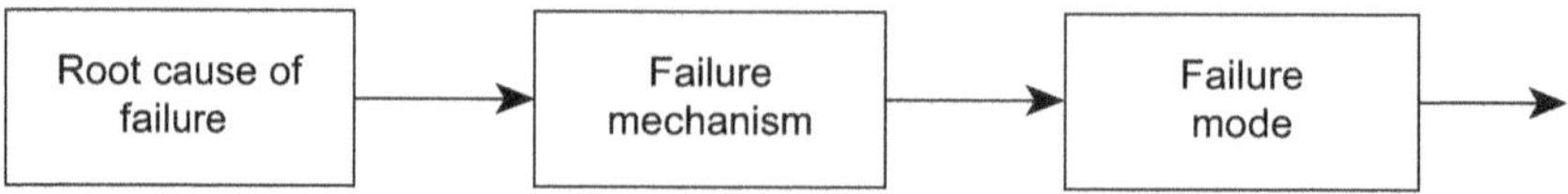

Figure 3.2   Relationship between root cause and failure mechanism

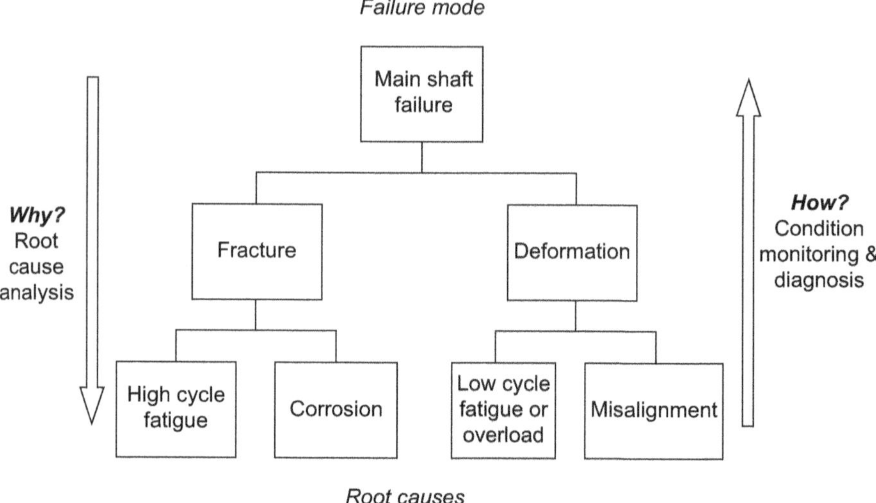

Figure 3.3   Relationship between failure mode and root cause for a WT main shaft failure

The importance of this relationship is that we can generally obtain good evidence of failure location, from which we can infer the failure mode, but for O&M purposes it is much more valuable to identify the root cause, which can be tracked by an operator or a maintainer to predict the progress of the incipient failure. This knowledge becomes invaluable to plan maintenance and reduce downtime. Figure 3.3 shows that monitoring data is a key ingredient to that tracking process.

3.6   Reliability field data

Once a WT taxonomy has been defined and the parts of the WT are named in a standardised way, data can be collected on WT reliability. A number of surveys of WT reliability exist in the public domain including the following:

i. Windstats surveys in Denmark and Germany [7], termed WSDK and WSD, respectively, containing data on failure rates fixed and variable speed WTs with geared or direct drives over 25 years of operation.

ii. Various Swedish and Finnish surveys mentioned in Reference 2.
iii. LWK survey in Germany [11] containing data on failure rates from fixed and variable speed WTs with geared or direct drives over 15 years of operation from 5800 WT years.
iv. WMEP survey in Germany [6] containing data on failure rates from fixed and variable speed WTs with geared or direct drives over 15 years of operation from 15,400 WT years.
v. ReliaWind survey in Europe [7, 8] of 450 wind farm months of data, comprising around 350 onshore variable speed WTs with geared drives operating for varying lengths of time ≤ 4 years, in the form of 35,000 downtime events each tagged within the standard taxonomy described above.

In general, the data from (i) WSD and WSDK above do not segregate failures between different types of WT or into different WT assemblies, whereas data from (ii), LWK (iii) and WMEP (iv) do. In addition, data from ReliaWind (v) subdivides the data from non-specific types of WTs into assemblies, sub-assemblies and some components, as prescribed by the taxonomy described in Section 3.4. Therefore, the data sources can be viewed as more detailed as one progresses down the list, with the exception that the ReliaWind data do not identify individual WT types to preserve confidentiality, whereas WMEP and LWK data do identify individual WT types.

At the date of writing very little field data exist in the public domain for offshore wind farms, although there are a number of reports published from the early publicly funded projects in Europe, see Reference 4.

3.7   Comparative analysis of that data

The simplest comparison of onshore WT reliability results has been done by the author in Reference 1 based on WMEP and LWK mixed WT data, and an extracted example is shown from LWK data in Figure 3.4.

Comparisons have also been published in References 2, 12 and 13.

Figure 3.3 shows how, in general, failure rates for stoppages >24 hours seem to be increasing with increasing WT rating.

The results of Figure 3.4 show that WT electrical sub-assemblies appear to have the higher failure rates but the highest downtimes are in the drive train due to the blades, gearbox and generator sub-assemblies. From the failure rates this is clearly not due to their intrinsic design weakness but rather the complexity of changing them in the field, entailing the use of cranes and the need for prior planning.

It is also interesting to note from Figure 3.5 the differences between downtime recorded by the two surveys. LWK represented the total downtime, whereas WMEP tried to record *MTTR* itself. *MTTR* is shorter than the downtime, as shown in (1.3), generally confirmed by Figure 3.5.

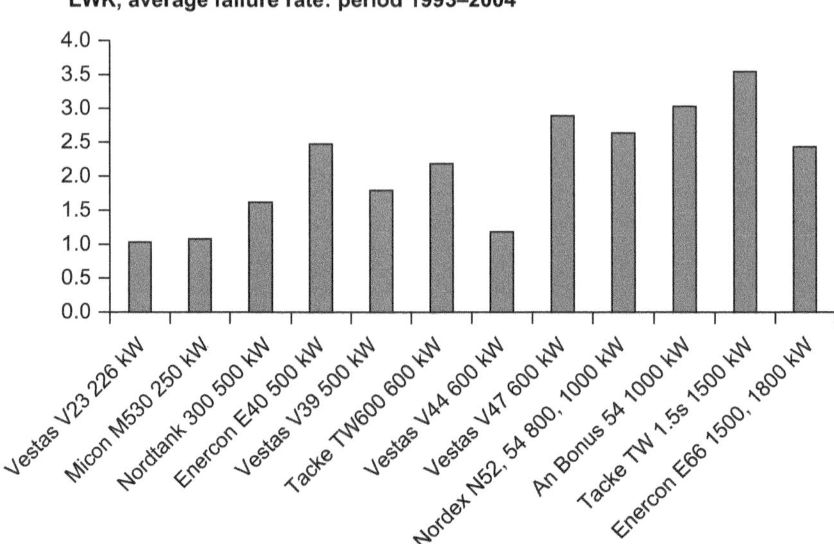

*Figure 3.4   Variation of WT failure rates with rating from the 5800 turbine-year
LWK survey between 1993 to 2004 [Source: [11]]*

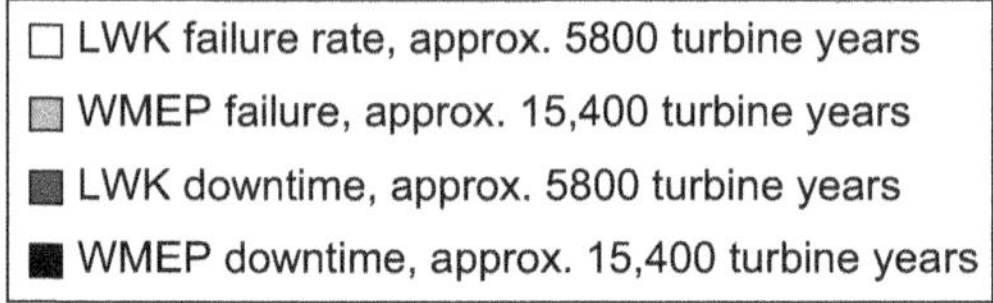

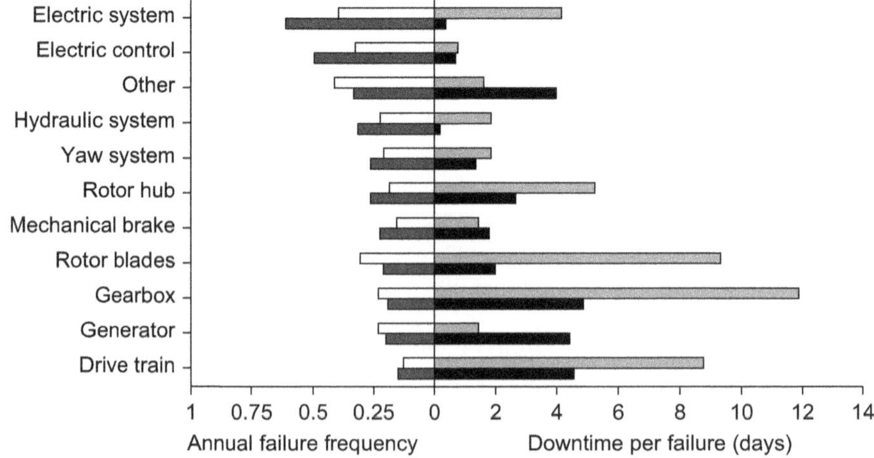

*Figure 3.5   WT sub-assembly failure rate and downtime per failure, the 20,000
turbine-year LWK and WMEP surveys, 1991–2004 [Source: [6, 11]]*

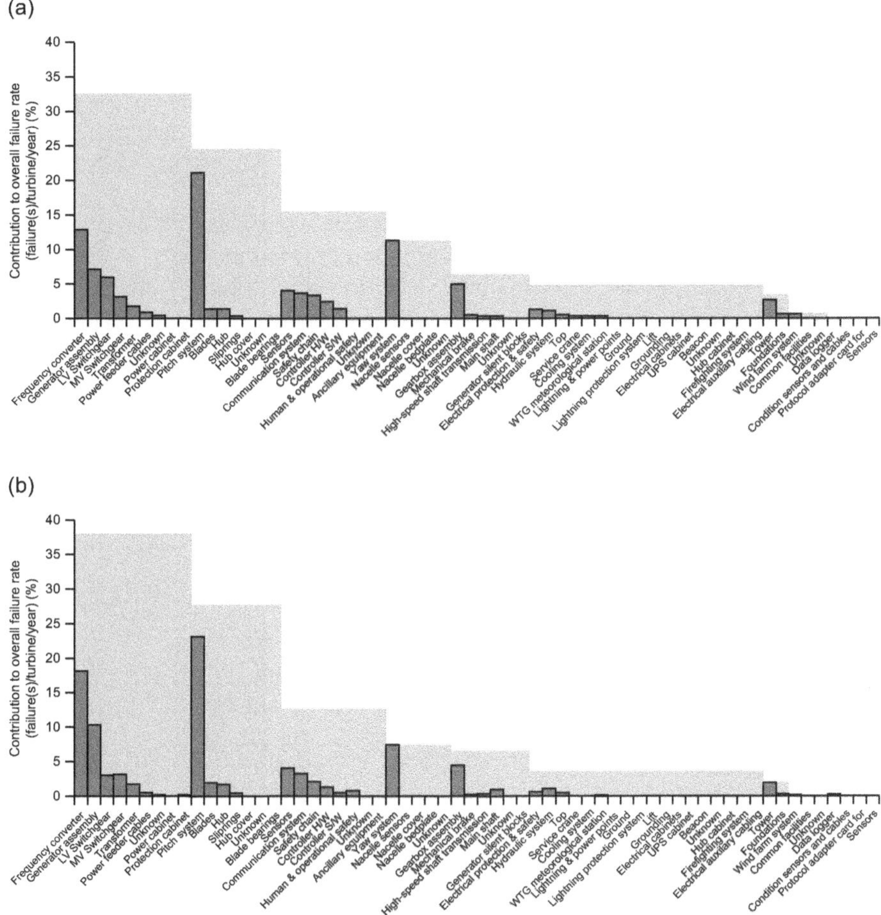

Figure 3.6   *WT sub-assembly reliability information from the 1400 turbine years ReliaWind survey, 2004–2010. (a) Sub-assembly failure rate distribution; (b) sub-assembly downtime distribution [Source: [12]]*

Figure 3.6 shows the results from the more recent ReliaWind survey that has a much more detailed breakdown of WT sub-assemblies and data collected for stoppages >1 hour. The failure rate lessons from ReliaWind (Figure 3.6(a)) are rather similar to the public domain surveys, but the downtime structure (Figure 3.6(b)) is different showing a much greater emphasis on the rotor and power modules because it is believed these newer WTs have not experienced any major gearbox, generator or blade failure to date in service.

3.8   Current reliability knowledge

On the basis of the above results, our current knowledge of onshore WT reliability is that:

- WT failure rates are generally falling with time, so the industry is producing more reliable WTs as time progresses.
- Failure rates of 1–3 failure(s)/turbine/year for stoppages ≥24 hours are common onshore, depending on the definition of a failure.
- Offshore, a failure rate of 0.5 failure(s)/turbine/year is likely to be necessary, where planned maintenance visits need to be kept at or below 1 per year, if possible.
- Failure rates vary with WT configuration but there is, as yet, no clear advantage in any one technology. The impression is given that any technology can achieve a reasonable reliability provided it has had sufficient operational experience and competent maintenance.
- Failure rates of WTs generally rise with WT size. This can be ascribed to rapid increases in WT design sizes over the last 15 years and their increasing complexity.
- WT sub-assemblies with the highest failure rates have been shown from public domain surveys to be, in descending order of significance:
    - Rotor pitch system
    - Converter (i.e. electrical control, electronics, inverter)
    - Electrical system
    - Rotor blades
    - Generator
    - Hydraulics
    - Gearbox
- Sub-assemblies with the highest downtimes have been shown to be, in descending order of significance:
    - Gearbox
    - Generator
    - Rotor blades
    - Pitch system
    - Converter (i.e. electrical control, electronics, inverter)
    - Electrical system
    - Hydraulics

The relative standing of these lists will vary with WT type and configuration and may be altered by time as a result of WT O&M and asset management strategies.

A recent study [13] has shown that onshore 75% of the faults cause 5% of the downtime, whereas 25% of the faults cause 95% of the downtime. Downtime onshore is dominated by a few large faults, many associated with gearboxes, generators and blades, requiring complex and costly replacement procedures.

The 75% of faults causing 5% of the downtime are mostly associated with the electrical plant, the converter, electric pitch systems, control equipment and switchgear, whose defects are relatively easy to fix in an onshore environment. It is known that a large proportion of WT alarms originate in the electrical systems.

The cost of offshore operations of WTs is likely to be profoundly affected by these figures. It is likely that the failure rates offshore will be similar to onshore but that downtimes will be hugely affected by the location of the offshore wind farm and its accessibility, greatly increasing the 5% of onshore downtime arising from 75% of faults.

3.9   Current failure mode knowledge

The ReliaWind work, presented in Figure 3.6, determined the six least reliable sub-assemblies in a 1400 turbine-year survey, summarised in descending unreliability as follows:

- Pitch mechanism, electric or hydraulic
- Power electronic converter
- Yaw system
- Control system
- Generator
- Gearbox

The ReliaWind project also conducted a failure modes and effects analysis of the WT type covered by the survey and this revealed the most important failure modes identified in those six sub-assemblies, as set out in Table 3.1.

The unreliable sub-assemblies have been identified objectively from measured data, whereas the failure modes were identified subjectively by ReliaWind partners.

3.10   Linkage between failure mode and root cause

The failure information in Figures 3.5 and 3.6 shows the location of failures, while Table 3.1 identifies failure modes. However, to raise reliability it is necessary to identify and if possible eliminate the root cause. The linkage between those two, described in Figure 3.2, depends upon the sequence shown in Figure 3.7.

Because of the distributed nature of wind power and the relatively low rating of individual WTs, it is rare for the WT OEM or operator to perform a root cause analysis on failures. Therefore, the knowledge of root cause must be built up in the industry by relying on the monitoring available from the wind far, a topic that will be developed in Chapter 7. It is important to note from Figure 3.7 that the weather plays a significant role in wind power, not only as the resource for energy conversion but also as a root cause for failure. This is developed and discussed in Chapter 15, Appendix 6.

Table 3.1 Unreliable sub-assemblies and failure modes identified through an FMEA (Failure Modes and Effects Analysis) in the ReliaWind project [Source: [10]]

Sub-system/ Assembly	Failure mode 1	Failure mode 2	Failure mode 3	Failure mode 4	Failure mode 5
Pitch system	Electrical (5 out of 13)	Battery failure	Pitch motor failure	Pitch motor converter failure	Pitch bearing failure
	Hydraulic (5 out of 5)	Internal leakage of proportional valve	Internal leakage of solenoid valve	Hydraulic cylinder leakage	Position sensor degraded or no signal
Frequency converter (5 out of 18)	Generator- or grid-side inverter failure	Loss of generator speed signal	Crowbar failure	Converter cooling failure	Control board failure
Yaw system (5 out of 5)	Yaw gearbox and pinion lubrication out of specification	Degraded wind direction signal	Degraded guiding element function	Degraded hydraulic cylinder function	Brake operation valve does not operate
Control system (5 out of 5)	Temperature sensor module malfunction	PLC analogue input malfunction	PLC analogue output malfunction	PLC digital input malfunction	PLC In line controller malfunction
Generator assembly (5 out of 11)	Worn slip ring brushes	Stator winding temperature sensor failure	Encoder failure	Bearing failure	External fan failure
Gearbox assembly (5 out of 5)	Planetary gear failure	High-speed shaft bearing failure	Intermediate shaft bearing failure	Planetary bearing failure	Lubrication system malfunction

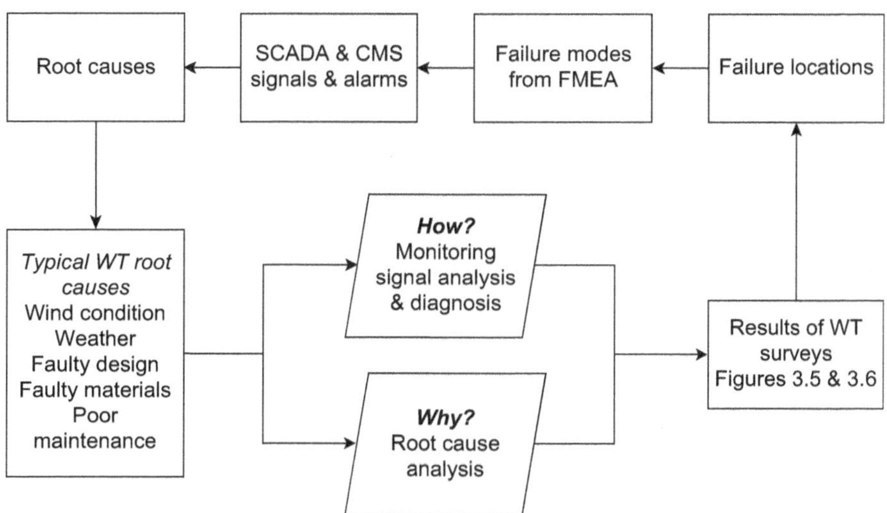

Figure 3.7   Linkage between WT failure location, failure mode and root cause

3.11   Summary

This chapter has demonstrated how WT reliability data can be used to benchmark WT performance for organising and planning future operations and maintenance, particularly offshore. Data need to be collected carefully and the taxonomy of WTs and wind farms must be defined in a common way, for which standards exist in other industries and are being prepared for the wind industry. It is also clear that the definition of failures, or indeed stoppages, need to be standardised to ensure that data can be compared in a useful engineering and management way.

Data are available in the public domain and give clear indications of the major reliability problem areas within the WT taxonomy. Failure rates of 1–3 failure(s)/turbine/year are common onshore for stoppages of $\geq$24 hours. Onshore, 75% of the faults cause 5% of the downtime, whereas 25% of the faults cause 95% of the downtime. It is likely that the figure of 5% of downtime, due to 75% of faults, will rise due to increased access times offshore and this will be due in large part to relatively minor faults that onshore were repaired by short and easy to arrange visits to site.

Failure rates of 0.5 failure(s)/turbine/year would be desirable offshore but are nowhere near this level.

From a limited survey, the chapter has finally shown, in Figure 3.6, the least reliable sub-assemblies in modern WTs, the failure modes causing that unreliability and the linkage between those failures and their root causes.

3.12   References

[1]   Tavner P.J., Xiang J.P., Spinato F. 'Reliability analysis for wind turbines'. *Wind Energy*. 2006;**10**(1):1–18

[2]   Ribrant J., Bertling L. 'Survey of failures in wind power systems with focus on Swedish wind power plants during 1997–2005'. *IEEE Transactions on Energy Conversion*. 2007;**22**(1):167–73

[3]   Spinato F. *The Reliability of Wind Turbines*. Doctoral thesis, Durham University; 2008

[4]   Feng Y., Tavner P. J., Long H. 'Early experiences with UK round 1 offshore wind farms'. *Proceedings of Institution of Civil Engineers*, Energy 2010; **163**(EN4):167–81

[5]   EN ISO 14224:2006, Petroleum, petrochemical and natural gas industries – collection and exchange of reliability and maintenance data for equipment

[6]   Faulstich S., Durstewitz M., Hahn B., Knorr K., Rohrig K. Windenergie Report, Institut für solare Energieversorgungstechnik, Kassel, Deutschland, 2008

[7]   *Windstats (WSD & WSDK) quarterly newsletter, part of WindPower Weekly, Denmark*. Available from http://www.windstats.com [Last accessed 8 February 2010]

[8]   Wilkinson M.R., Hendriks B., Spinato F., Gomez E., Bulacio H., Roca J., *et al.* 'Measuring wind turbine reliability: results of the ReliaWind project, Scientific Track'. *Proceedings of European Wind Energy Conference EWEA2011*. Brussels, Belgium: European Wind Energy Association; 2011

[9]   VGB PowerTech. *Guideline, Reference Designation System for Power Plants, RDS-PP, Application Explanation for Wind Power Plants, VGB-B 116 D2*. 1st edn. Essen: VGB PowerTech e.V.; 2007

[10]   ReliaWind. *Deliverable D.2.0.4a-Report, Whole system reliability model*. 2011. Available from http://www.reliawind.eu/files/file-inline/110318_Reliawind_DeliverableD.2.0.4aWhole_SystemReliabilityModel_Summary.pdf [Accessed 10 May 2012]

[11]   *Landwirtschaftskammer (LWK). Schleswig-Holstein, Germany*. Available from http://www.lwksh.de/cms/index.php?id¼1743 [Last accessed 8 February 2010]

[12]   Wilkinson M.R., Hendriks B., Spinato F., Gomez E., Bulacio H., Roca J., *et al.* 'Methodology and results of the ReliaWind reliability field study'. *Proceedings of European Wind Energy Conference*, EWEC2010. Brussels, Belgium: European Wind Energy Association; 2010

[13]   Faulstich S., Hahn B., Tavner P. J. 'Wind turbine downtime and its importance for offshore deployment'. *Wind Energy*. 2011;**14**(3):327–37

Chapter 4

Effects of wind turbine configuration on reliability

4.1 Modern wind turbine configurations

Section 1.1 showed that modern electric power HAWTs have evolved over the last 80 years not only in rating but also in the number and variety of configurations as follows:

- Upwind or downwind WT rotors
- Two- or three-blade rotors
- Fixed speed or variable speed rotors
- Stall regulated or pitch regulated
- Direct drive or geared drive

More recently they have standardised towards three-blade, upwind, pitch-regulated rotors, growing in size as exemplified by Figure 4.1.

The variations in WTs are now more concentrated on the drive train itself and the electrical arrangements of these configurations, and these features affect the turbine performance and therefore its reliability. So when considering reliability, a clear understanding of the configuration and its strengths and weaknesses are very important. Some parts of the industry have perceived certain configurations to be more reliable than others but as yet no clear measured data seem to point in that direction. In fact, recent experience has emphasised that any configuration can achieve reliability provided that the component sub-assemblies are well manufactured, well installed and well maintained.

Figure 4.2, based on the nomenclature used in Reference 2, summarises the main drive train configurations currently in use in the industry as follows:

- Type A for fixed- or dual-speed, stall-regulated WTs with a geared drive low-voltage (LV) squirrel cage induction generator (SCIG) connected directly to the medium-voltage (MV) grid through a transformer, with power factor correction and a soft starter to reduce synchronisation inrush current.
- Type B for fixed- or dual-speed, stall-regulated or variable-speed, controlled-stall-regulated WTs with a geared drive LV wound rotor induction generator (WRIG) with variable rotor resistance connected directly to the MV grid

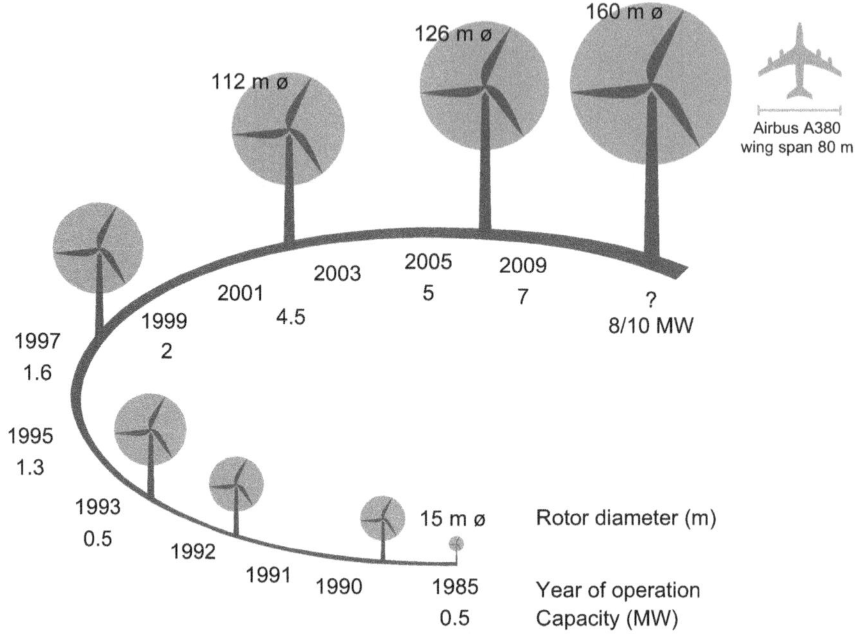

Figure 4.1   Growth in size of commercial wind designs 1985–2009 [Source: EWEA [1]]

through a transformer with power factor correction and a soft starter to reduce synchronisation inrush current.

- Type C for variable-speed, variable-pitch WTs with a geared drive LV WRIG and partially rated, four-quadrant converter connected to the WRIG rotor, whose stator is connected to the MV grid through a transformer. This is the so-called doubly fed induction generator (DFIG) scheme that is the most widely fitted in the wind industry for WTs $\geq$ 1.5 MW.
- Type D for variable-speed, variable-pitch WTs with a direct drive LV wound rotor synchronous generator with exciter (WRSGE), permanent magnet synchronous generator (PMSG) or SCIG with a fully rated four-quadrant converter connected to the stator, which is connected to the MV grid through a transformer.

4.2   WT configuration taxonomy

4.2.1   General

Chapter 3 has shown that the reliability of large modern onshore WTs is improving but the wind industry must have a clear understanding of the factors driving this reliability to face the economic challenges of offshore installations, where the wind energy harvest is greater but the conditions are more inclement. It will be necessary

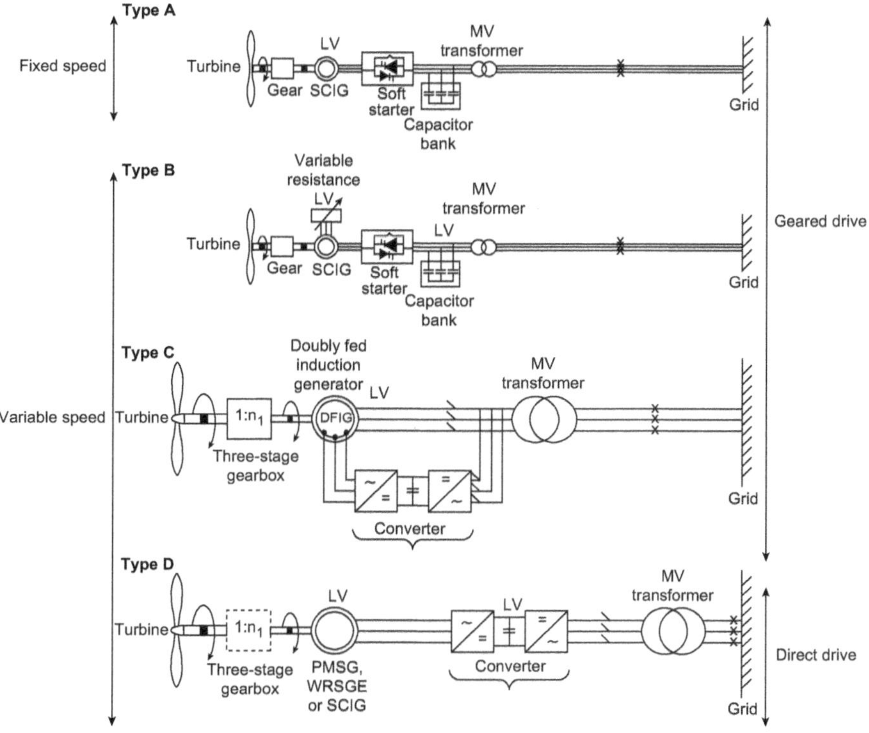

Figure 4.2   Summary of main electrical configurations for current WT drive trains

to increase reliability further, because access to those WTs will be more limited. Chapter 2 showed that a failure rate of 1–3 failure(s)/turbine/year are common onshore, and some would argue that real failure rates are much higher than that if all stoppages are taken into account. Offshore, a failure rate of 0.5 failure(s)/turbine/ year is likely to be necessary, where planned maintenance visits need to be kept at or below 1 per year if availability and low cost of O&M and energy are to be achieved.

This section considers the unreliability or failure intensity function, $\lambda(t)$, of WT sub-assemblies rather than the wider issue of availability and capacity factor (CF) because reliability depends primarily on WT construction and is intrinsically pre-dictable. On the other hand, availability, yearly production and CF depend not only on reliability but also more strongly on wind conditions and the consequences of faults, which in turn depend on turbine location, access logistics and maintenance regime, not primarily to the WT construction. This section carries forward the ana-lysis of Section 3.5 on public domain data, paying particular attention to vital WT sub-assemblies, the gearbox, generator and power electronic converter. The foun-dation of these analyses has been the population of WTs of known model and design covered by the LWK survey (see Reference 11 of Chapter 3). They will show striking differences between the reliability characteristics of the selected sub-assemblies over the period. Some of the results can be related to experience with such sub-assemblies outside the wind industry. Considerable interest has also been shown in the industry about differences in cost and performance achieved by different WT architectures, see for example Reference 3, but reliability information was lacking. The analysis here sheds light on the effect of WT configuration on reliability and identifies specific reliability behaviours of selected sub-assemblies, where work could be done to improve overall WT.

4.2.2   Concepts and configurations

As the technology of modern WTs has matured, the construction has become standardised around the three-bladed, upwind, variable speed concept. But within this concept there are different architectures and Types C and D in Figure 4.2 show two, as follows:

- Geared WTs with a gearbox, a high-speed asynchronous generator and a partially rated converter (DFIG)
- Direct drive WTs with no gearbox but a specialised direct drive, low-speed synchronous generator and fully rated converter

The anticipated benefit of the geared concept is that it uses a more standardised, high-speed generator and a partially rated converter, thereby saving cost as shown in Reference 3. An anticipated benefit of the direct drive concept is that by avoiding the use of the gearbox it should prove to be more reliable but there are other potential benefits, for example lower losses in low wind. There are also a number of control configurations that need consideration and these are listed in Table 4.1. This chapter will investigate the reliability of a number of these turbine concepts where a concept means the sum of the WT architecture and control configuration.

Table 4.1   WT control concepts considered in this chapter

Speed control	Pitch control	Power control	WT models considered in this section
Fixed or dual speed	None	Passive stall regulation geared drive train with SCIG	NEG Micon, M530, Tacke TW600
Fixed speed	Yes, pitch to stall	Active stall regulation geared drive train with SCIG	Vestas V27, Nordex N52/54
Limited variable speed	Yes	Geared drive train with WRIG control using variable rotor resistance	Vestas V39
Variable speed	Yes	Geared drive train with DFIG control using partially rated converter	Tacke TW1500, Bonus 1 MW, 54
Variable speed	Yes	Direct drive train with synchronous generator control using fully rated converter	Enercon E40, E66

4.2.3   Sub-assemblies

To understand WT drive train configuration reliability we need to break down the WT into more detail than Figure 2.1 using the nomenclature in Section 3.4:

- System, the whole WT
- Sub-systems of the WT, such as the drive train, consisting of rotor hub, shaft, bearing, gearbox, couplings and generator
- Assemblies, such as the gearbox
- Sub-assemblies, such as the high-speed shaft of the gearbox
- Components, such as the high-speed coupling of the gearbox

This chapter focuses on sub-assemblies recorded in the surveys WSDK, WSD and LWK, and the sub-assembly breakdown is shown in Figure 3.5. The terminology used by these surveys was not consistent and it has been necessary to aggregate sub-assemblies as shown in Table 4.2.

Table 4.2   WT sub-assemblies considered in this chapter

This chapter	WSD	WSDK	LWK
Rotor	Rotor	Blades, hub	Blades
Air brake	Air brake	Air brake	Rotor brake
Mechanical brake	Mechanical brake	Mechanical brake	Brake
Main shaft	Main shaft, bearings	Axle, bearing, coupling	Shaft, bearings
Gearbox	Gearbox	Gearbox	Gearbox
Generator	Generator	Generator	Generator
Yaw system	Yaw system	Yaw system	Yaw system
Converter	Electrical control	Electrical control	Electronics, inverter
Hydraulics	Hydraulics	Hydraulics	Hydraulics
Electrical system	Electrical system	Grid	Electrics
Pitch control	Pitch adjustment	Mechanical control	Pitch mechanism
Other	Anemometry, sensors, other	Other	Anemometry, sensors, other

4.2.4   Populations and operating experience

WSD, WSDK and LWK data (see Reference 3 in Chapter 2 and Reference 11 in Chapter 3) were collected by operators on hand-written or computer-written report sheets, rather than generated automatically, and the data have some limitations, as follows:

i. They gather the failures in a given period for each turbine and sub-assembly within the population without giving details of failure modes.
ii. The periods of data collection differ for each population as follows: WSDK monthly, WSD quarterly, LWK annually.
iii. These periods have affected the results presented.

iv. There are other differences between the populations as follows:

- o WSDK is a large mixed population decreasing in WT numbers (2345–851 over the period), with turbines of average age of 14 years, mostly of stall-regulated configuration. Their technology is consolidated as confirmed by their failure intensities, approaching a constant average failure rate. The failures of individual turbine models cannot be distinguished in this data.
- o WSD is a larger mixed population growing in number (1295–4285 over the period) and includes larger turbines, with an average age of 3 years, including a variety of turbine models with different control configurations but their failure intensities also approaching a constant value, although at a faster rate than WSDK. The failures of individual turbine models again cannot be distinguished in this data.
- o LWK is a smaller, segregated, more static population in number (158–643 over period) and includes larger turbines of average age up to 15 years, with fixed and variable speed configurations, both with geared and a significant number with the direct drive concept. The failures of individual turbine models can be distinguished in this data.

4.2.5   Industrial reliability data for sub-assemblies

Some WT sub-assemblies, such as the rotor and pitch control, are specialised for the wind power application. But some, such as the gearbox, generator and converter can be found in similar form, albeit in different sizes and designs, in other power conversion machinery. The usefulness to the industry of reliability figures presented in this chapter is enhanced by comparing them to values from other industries, as tabulated in Table 4.3.

Table 4.3   Reliability of generators, gearboxes and converters from industrial experience

Sub-assembly	Failure rate (failures/ sub-assembly/year)	MTBF (hour)	Source
Generator	0.0315–0.0707	123,900–278,000	Tavner [4, 5] and IEEE Gold Book [6]
Gearbox	0.1550	56,500	Knowles
Converter	0.0450–0.2000	43,800–195,000	Spinato (Reference 7 of Chapter 2)

4.3   Reliability analysis assuming constant failure rate

Previous work by the authors of Reference 3 of Chapter 1 concentrated on the average WT failure rate, assuming the systems were at the bottom of the bathtub (Figure 2.4). This showed for WSD, WSDK (Figure 1.2), the overall trend in WT failure intensities against calendar time since the days of the early expansion of

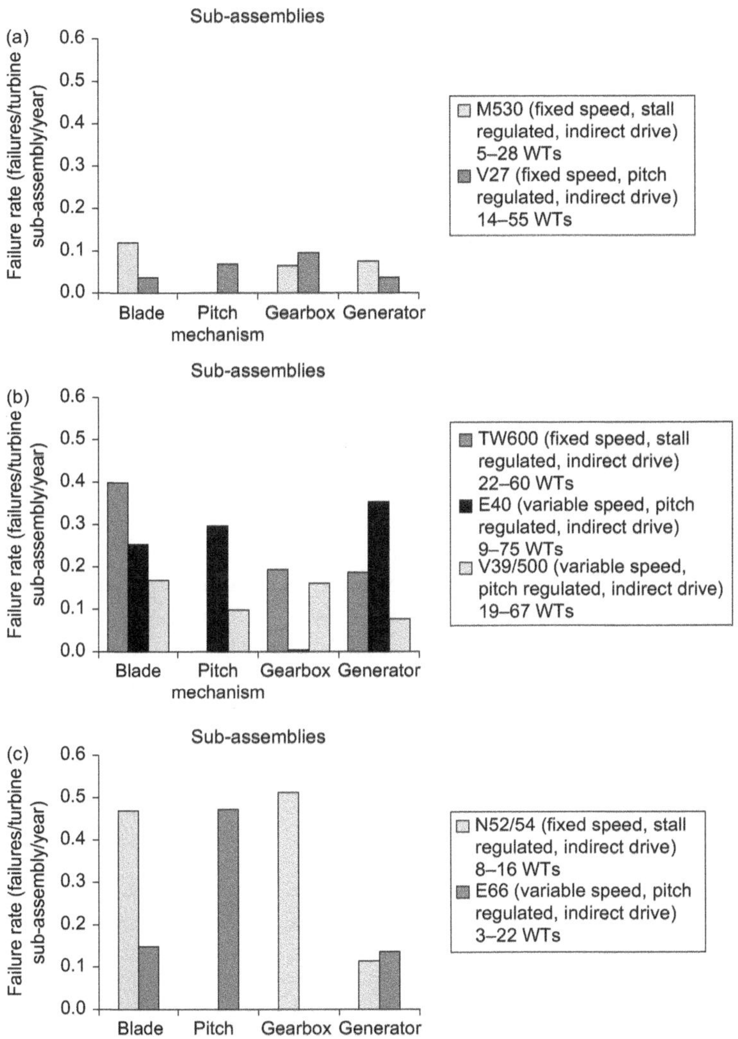

Figure 4.3   LWK failure intensity distributions, as in Figure 3.5, focussing on blades, pitch, gearbox and generator: (a) 300 kW fixed speed, geared turbines with pitch- or stall-regulated control; (b) 600 kW fixed speed, geared, stall-regulated or limited variable speed pitch-regulated turbines or variable speed direct drive pitch-regulated turbines; (c) 1 MW variable speed geared, pitch-regulated, turbines or variable speed direct drive, pitch-regulated turbines. Stall-regulated turbines on left, variable speed, pitch-regulated turbines on right [Source: Reference 6 of Chapter 2]

wind power in California in the early 1980s. The results of the LWK survey have been added to Figure 1.2 with the measured failure rates from other mature power generation sources, largely extracted from IEEE sources [5], showing that WT reliability is becoming better than some other generation sources, notably diesel generator sets. However, this graph needs to be treated with caution for the following reasons:

- The WT data are taken from mixed and changing WT populations. Because the ratings of newly introduced WTs are increasing and their failure rates are generally rising, the averaging implicit in the HPP process tends to underestimate the failure rates of these newer, larger, more complex WTs, at least during the early failures period.
- The other, mature power generation source, failure data came from historic surveys of limited size, which do not represent the reliability improvement to be studied in this chapter but which is also inherent in those sources.

The relative unreliability of WT sub-assemblies can also be extracted from the WSD and LWK data as shown in Figure 3.5, where the assumed constant failure rates of 11 major turbine sub-assemblies have been compared. The LWK population has a higher consistency in terms of technology throughout the period, as it is an installed fleet that has remained relatively unchanged. However, the LWK population is much smaller than the WSD populations. Figure 4 reveals interesting information:

- Overall failure rates in Danish turbines are lower than German turbines, as seen in Figure 1.2. This was attributed in Reference 3 of Chapter 1 to the greater age, smaller size and simpler technology of the Danish turbines resulting in a higher overall reliability.
- Figure 3.5 shows that the failure rates of sub-assemblies in the two German populations, WSD and LWK, are remarkably similar and have more in common with one another than with the WSDK data. This consistency supports the validity of the two German surveys despite their different sizes.
- The results of Figure 3.5 show that the sub-assemblies with the highest failure rates are in descending order of significance:
  - Electrical system
  - Rotor (i.e. blades and hub)
  - Converter (i.e. electrical control, electronics, inverter)
  - Generator
  - Hydraulics
  - Gearbox

Similar results have been reported from Sweden (see Reference 3 of Chapter 1) and from a different survey in Germany, WMEP (see Reference 6 of Chapter 3).

The failure rates obtained for WT sub-assemblies will also be compared in this chapter with those obtained from industry (see Table 4.3).

Figure 3.5 considers failure rate only and not failure severity. However, LWK data record the downtime or *MTTR* of different sub-assembly failures and this is shown in Figure 3.5. Here the effects of electrical system, generator and gearbox failures are more apparent, in particular the dominance of the gearbox *MTTR*. It is suggested that this is the main reason for the industry's focus on gearbox failures. Again, similar results have also been obtained in Sweden (see Reference 3 of Chapter 1).

4.4   Analysis of turbine concepts

4.4.1   Comparison of concepts

We now consider the failure rates of individual sub-assemblies most at risk. The LWK data allow turbine models to be grouped according to size and concept. Figure 3.4 summarised the failure rates over 11 years for 12 WT models in the LWK population, as listed in Table 4.1. This shows the general trend of failure rate rising with turbine rating, reaffirming a conclusion of Reference 2 of Chapter 1. The next analysis repeats the approach of Figure 3.5, comparing sub-assembly failure rates for selected LWK turbine models, concentrating on drive train sub-assemblies. This is shown in Figure 4.3, which is segregated by turbine concept and control configuration, see the third column of Table 4.1.

The figure shows the relationship between failure rates of blades, pitch mechanism, gearbox and generator as turbine concepts and control configurations change.

With fixed speed, stall-regulated turbines, a significant number of failures are concentrated in the blades and gearbox. With the introduction of variable speed, pitch-regulated machines, the pitch mechanism now appears as a failure mode, as expected.

However, the introduction of the pitch mechanism reduces the blade and generator failure rates, see Figure 4.3(a) for smaller WTs. This is confirmed for larger WTs in Figure 4.3(b) where blade, generator and gearbox failure rates reduce, with the exception of the E40, direct drive WT, where the generator failure rate was high. This will be discussed in Section 4.4.2. The reduction in blade failures is even more noticeable with the larger E66 direct drive WT in Figure 4.3(c).

In other words, the technological advance of variable speed and pitch control not only confers energy extraction and noise reduction improvements but also, despite introducing other failure modes, can improve WT reliability with time.

4.4.2   Reliability of sub-assemblies

4.4.2.1   General

The failure data collected exhibit a variation with time and can be represented by an NHPP, see Section 2.4.5. This section will now use reliability growth analysis, based on Figure 3.3 the PLP representation, a specific case of the NHPP, to analyse reliability time trends from the LWK population of WTs concentrating on three sub-assemblies identified above:

- Generator
- Gearbox
- Converter, that is, electrical control, electronics, inverters

These sub-assemblies have been chosen because they are crucial to WT operation and are central to the debate about turbine concept, in particular whether to employ direct drive or geared WTs.

The method of presentation is to plot the intensity function obtained from the LWK data against total time on test (*TTT*) of the sub-assembly, see Section 2.4.7. Plotted failure intensity points have been aggregated to comply with requirements for valid numbers of failures in an interval and the Crow-AMSAA model, as described in Section 2.4. On each graph, the failure rate of that sub-assembly in other industries taken from Table 4.3 is also shown, together with a time cursor to demonstrate the span in years of the data, as described in Section 2.4.7.

For these sub-assemblies, the PLP interpolation of data presented has been tested against two statistical criteria:

- Goodness of fit
- Null hypothesis of no reliability growth

Only results complying with those criteria have been presented. Sub-assemblies from specific WT models are selected here but the conclusions drawn below may be generalised to other WTs in the LWK population. These results have all been summarised in Reference [11].

4.4.2.2 Generators

Figure 4.4 shows the reliability of a number of LWK generators showing that failure intensities are generally falling, that is a PLP with $\beta < 1$, reflecting that reliability is improving. The early failures of the bathtub curve (Figure 2.4) can clearly be seen in these figures. Industrial generator reliability data, given in Table 4.4, is superimposed on the graphs and Figure 4.4 shows that both direct and geared drive WT generator reliabilities are not as good as these at the start of life. However, the demonstrated reliabilities, as defined in Figure 2.6, achieved by all except the E40 generator shows a good result when compared with the industrial failure rate. The failure intensities for both direct drive generators are higher than the failure intensities of their geared drive competitors. However, it is clear that the E66 generator is a considerable improvement on the E40 generator.

More recent information has come to light from a WT repair company [7] about WT generator failure rates compared to electrical machines in other industries. This confirmed that WT generators are not as reliable as similar-sized electrical machines in other industries, as seen in Figure 4.4, but throws more light on the location of WT generator failures, see Table 4.4 summarised in Figure 4.5. These confirm that the location of WT generator failures are not dissimilar from other electrical machines but are dominated by bearing, slip-ring and brush-gear

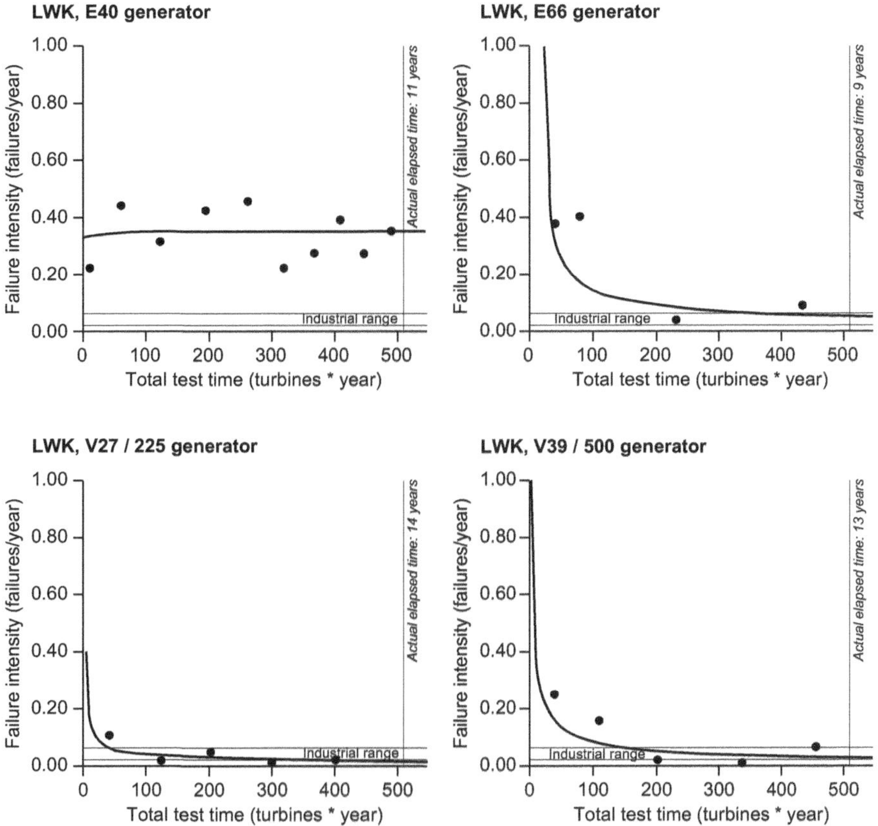

Figure 4.4   Variation between failure intensities of generator sub-assembly, in LWK population, using PLP model. Upper graphs: Low-speed direct drive, Fig 4.4(a); High-speed geared drive generators, Fig 4.4(b) [Source: Reference 6 of Chapter 2]

faults. This is not unexpected as the majority of large WT generators are currently DFIG.

Important questions are raised by these results as follows:

- Why is there such a large disparity between the reliabilities of direct and geared drive generators at the start of operational life?
- Why do the failure intensities of three generators improve with time?
- Why cannot the wind industry achieve, at the start of operational life, the respectable demonstrated reliabilities ultimately achieved?

These questions suggest, from this limited extract from LWK data, that generators deserve reliability attention from OEMs and operators if we are to achieve higher WT reliability, and this is discussed in Chapter 5.

Table 4.4   Distribution of failed sub-assemblies in electrical machines taken from literature

Surveys	IEEE large motor survey [8]	Motors in utility applications [9]	Motor survey offshore and petrochemical [10]	WT generator survey [7]		
Industry	**General**	**Utility applications**	**Offshore petrochemical**	**Wind generation**		
Types of machines, rating and voltage	Motors >150 kW generally MV and HV squirrel cage induction machines	Motors >75 kW generally MV and HV squirrel cage induction machines	Motors >11 kW generally MV and HV squirrel cage induction machines	Wind generators <1 MW, LV, 95% + wound rotor machines but with electronically controlled rotor voltage rather than collector rings with outboard electronics	Wind generators 1–2 MW, LV, mostly DFIGs	Wind generators >2 MW, LV, mostly DFIGs
No of failed machines in survey	360	1474	1637	196	507	297
Sub-assemblies						
Bearings	41%	41%	42%	21%	70%	58%
Cooling system	–	–	–	–	2%	–
Stator wedges	–	–	–	–	–	14%
Stator related	37%	36%	13%	24%	3%	15%
Rotor related	10%	9%	8%	50%	4%	4%
Collector or slip-rings	–	–	–	1%	16%	4%
Rotor leads	–	–	–	–	1%	4%
Other	12%	14%	37%	4%	4%	1%
Total	100%	100%	100%	100%	100%	100%

Source: [4]

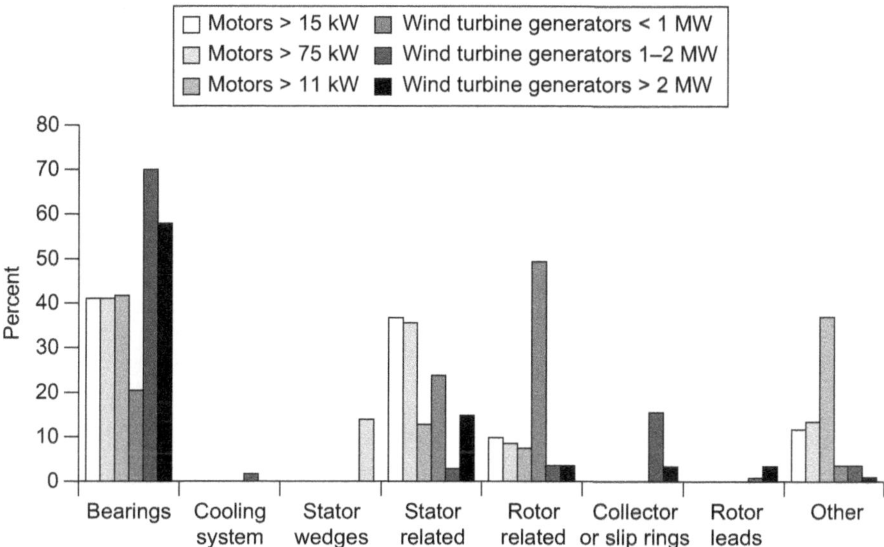

Figure 4.5   Location of failures in WT generators and other electrical machines [Source: [7]]

4.4.2.3 Gearboxes

Figure 4.6 shows the results for the reliability of a number of LWK gearboxes, which each show a remarkably similar form with rising failure intensities, which is a PLP with β from 1.2 to 1.8 (see Figure 2.2). That is, the deterioration or wear-out phase of the bathtub curve (Figure 2.4), suggesting steady mechanical wear, as one would expect. So WT gearboxes are a mature technology and machines are operating in the deterioration phase of the bathtub curve. Therefore, substantial improvements in designed reliability for these gearboxes are unlikely.

The reliability data for industrial gearboxes, given in Table 4.3, is an average from a number of sources and has been superimposed on the graphs. It shows that, from this limited extract from LWK data, reliabilities being obtained by these wind industry gearboxes are comparable with those obtained by other industries, apart from the Nordex 52/54 WT data.

4.4.2.4 Converters

The converter is a complex sub-assembly with a large number of components. There is difficulty in recording failures for converter sub-assemblies as operators may be unable to assign a turbine failure unequivocally to the converter because the sub-assembly is complex. This is in contrast to the generator or gearbox where this is usually straightforward. This means that we must be cautious in considering recorded converter failures.

To overcome this, we have aggregated the failures from inverter and electronics in the LWK survey (see Table 4.2), and the data have been plotted for specific turbines with the generic sub-assembly name, converter. Figure 4.7 gives reliability

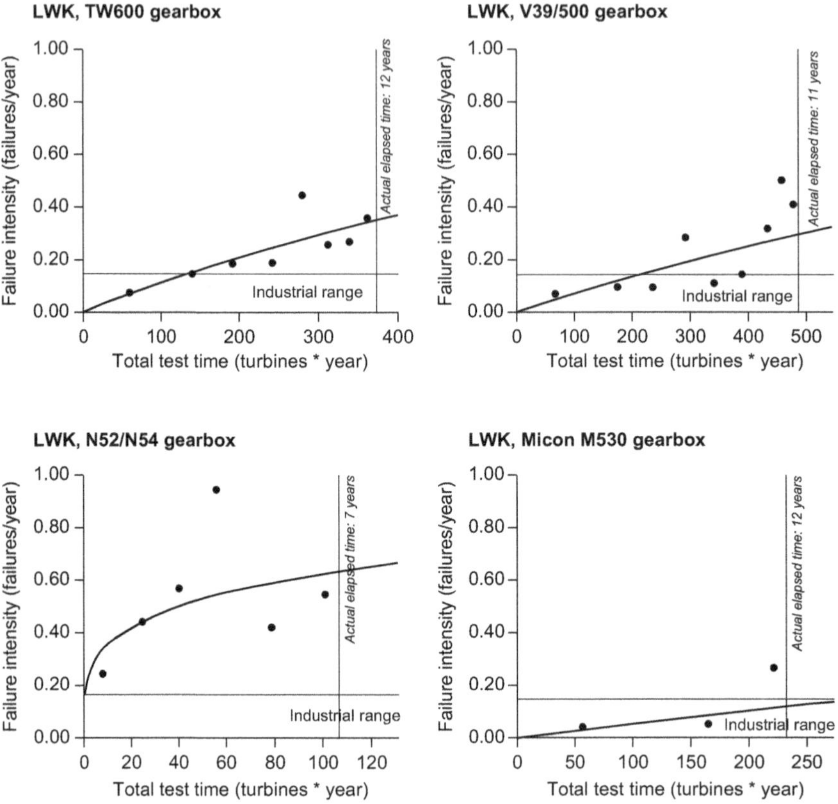

Figure 4.6   Variation in failure intensities of gearbox sub-assembly, in LWK population, using PLP model [Source: Reference 6 of Chapter 2]

results for three LWK converters. Again these exhibit the early part of the bathtub curve (Figure 2.4), but specifically the first curve shows elements of the full bathtub curve with early failures, intrinsic failures and wear-out. For two cases, the Enercon E40 and TW 1500, the results are similar to the generators, in that failure intensities are falling, that is a PLP with $\beta < 1$, reflecting reliability improvement. However, in the case of the Enercon E66 and E40 converters, the failure intensities improve with time but are nearly flat with $\beta = 1$. Industrial converter failure rate data in Table 4.2 range between 0.045 and 0.2 failure/sub-assembly/year. The lower limit arises from a specific analysis of relatively small converters (see Reference 7 of Chapter 2), but such a low value of failure rate cannot be applicable to the larger converters in WTs; therefore, an upper limit of 0.2 failure/sub-assembly/year is proposed.

More recent work has tracked the distribution of WT failure rates due to the converter, as shown in Table 4.5, where failures due to the converter are compared between different surveys. These show failure rates for converters ranging from 0.22 to 2.63 failures/unit/year, which should be compared to those shown in Figure 4.7.

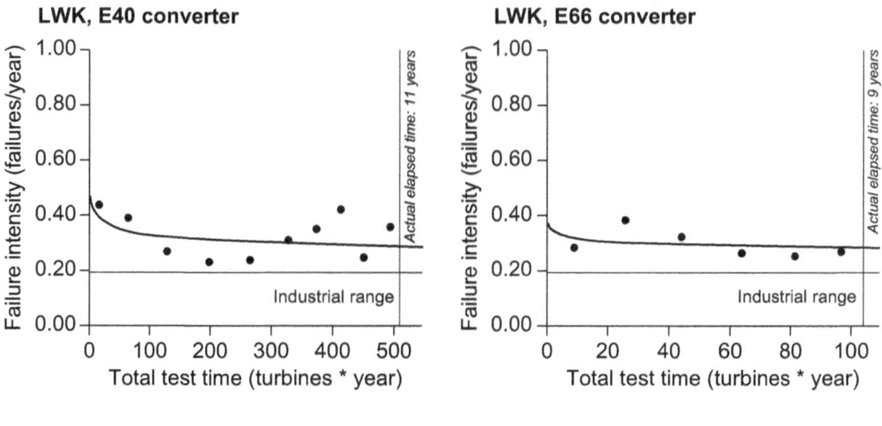

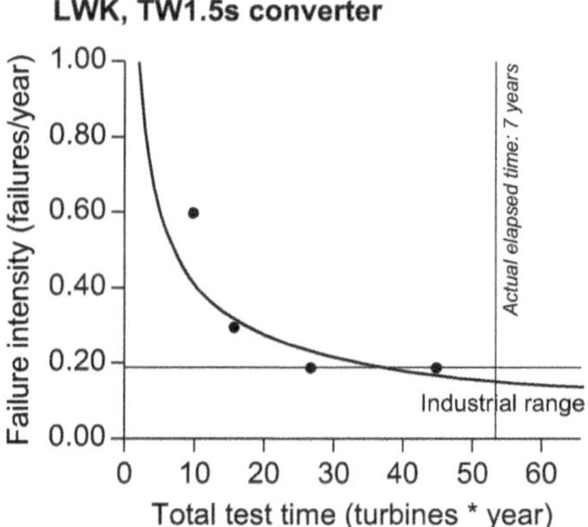

Figure 4.7   Variation in failure intensities of converter sub-assembly, in LWK population, using PLP model [Source: Reference 6 of Chapter 2]

It is important to point out here that the figures in Table 4.5 represent WT stoppages, ascribed by the operator as due to converter faults. These arise from the many alarm signals and trips that the converter produces. That does not locate the faults in the converter, which have been estimated in the lower rows of Table 4.5, based upon knowledge of converter sub-assembly reliabilities. It can be seen that the inverter bridge and DC link failures dominate converter failure rates and downtimes.

Despite its limitations, the data in Table 4.5 give a clear and consistent picture across a variety of surveys of the converter failure rate issue.

Important questions arise from these results as follows:

- Why do the failure intensities of converters improve with time?
- Why are the failure intensities considerably higher than values given for converters in normal industrial use?
- Why is not more attention being placed on reducing the high number of converter failures, perhaps by improving the alarm management and minimising the number of nuisance trips?

Table 4.5   *Distribution of failed converter sub-assemblies from various surveys*

	From WMEP_D data (Reference 6 of Chapter 3)	From LWK_D data (Reference 11 of Chapter 3)	From LWK_D (Reference 11 of Chapter 3)	From ReliaWind (Reference 12 of Chapter 3)		
Turbine years in the survey	209	1028	5719	679	366	1
Additional information	Large WTs	Total WTs	Total WTs	Specific data from WTs, with partially rated or Fully rated converter (E40, E66, Tacke 1.5s)	Specific data from WTs, about 2 MW with DFIG and partially rated converter	
	1998–2000	1989–2006	1993–2006	1993–2006	2007–2011	
	Failure rate (failures/unit/year)			Failure rate (failures/unit/year)	From FMEA failure rate (failures/unit/year)	
Whole WT	5.23	3.60	1.92	2.60	Not disclosed for confidentiality reasons	23.37
Converter total	1.00	0.45	0.22	0.32	Not disclosed for confidentiality reasons	2.63
Converter as % of WT	19.1%	12.4%	11.6%	12.2%	11.6%	11.3%

Estimated location of the faults

Converter control unit	0.070	0.031	0.016	0.022	—	0.184
Series contactor	0.090	0.040	0.020	0.028	—	0.237
Grid-side filter	0.030	0.013	0.007	0.009	—	0.079
Grid-side inverter	0.189	0.085	0.042	0.060	—	0.500
Pre-charge circuit	0.060	0.027	0.013	0.019	—	0.158
DC link capacitor	0.110	0.049	0.024	0.035	—	0.289
Chopper circuit	0.060	0.027	0.013	0.019	—	0.158
Generator-side inverter	0.189	0.085	0.042	0.060	—	0.500
Crow-bar circuit	0.060	0.027	0.013	0.019	—	0.158
Generator-side filter	0.030	0.013	0.007	0.009	—	0.079
Bypass contactor	0.090	0.040	0.020	0.028	—	0.237
Auxiliaries	0.025	0.011	0.006	0.008	—	0.066

These suggest, from this limited extract from LWK data, that converters deserve reliability attention from OEMs and operators if we are to achieve higher WT reliability, and this is discussed in Chapter 5.

4.5   Evaluation of current different WT configurations

In Reference 3, Polinder *et al.* evaluated five current 3 MW different WT drive configurations of which 4 are shown in Figure 4.8:

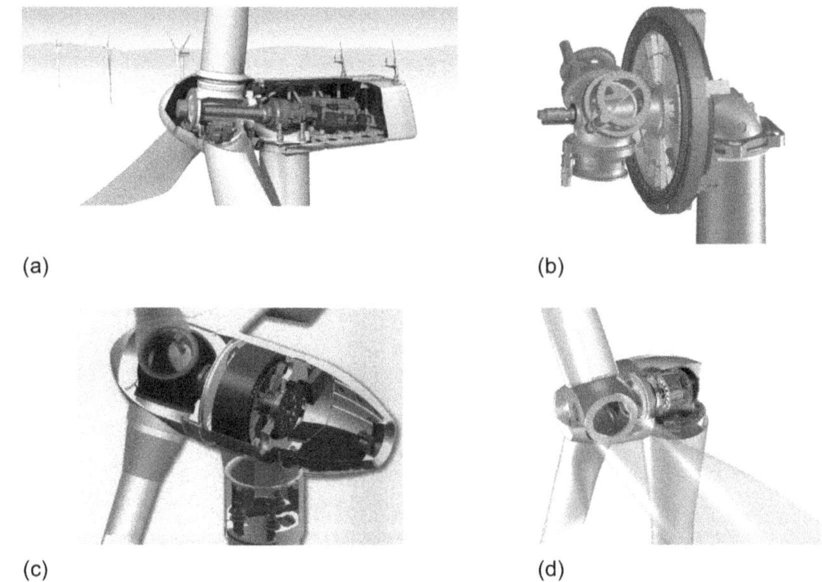

(a)

(b)

(c)

(d)

Figure 4.8   Pictures of different WT configurations evaluated. (a) Conventional geared DFIG3G; (b) conventional direct drive DDWRSGE; (c) permanent magnet direct drive DDPMG; (d) integral design PMG1G [Source: [3]]

- The indirect drive DFIG with three-stage gearbox and partially rated converter (DFIG3G), the turbine speed range being 3:1, therefore the converter rating is usually about one-third of that of the generator and gearbox
- The direct drive wound synchronous generator with electrical excitation and fully rated converter (DDWRSGE)
- The direct drive permanent magnet generator with fully rated converter (DDPMG)
- The semi-direct drive permanent magnet generator with a single-stage gearbox and fully rated converter (PMG1G)
- The semi-direct drive DFIG with single-stage gearbox and third-rated converter (DFIG1G)

Table 4.6   Evaluation of 3 MW drive train configurations with addition of reliability [Source: [3]]

	DFIG3G	DDWRSGE	DDPMG	PMG1G	DFIG1G
Annual energy yield (GWh)	7.73	7.88	8.04	7.84	7.80
Weight (kg)	5.3	45.1	24.1	6.1	11.4
Cost (euro)	1870	2117	1982	1883	1837
Estimated relative reliability (%)	90	70	80	100	80

The evaluation, Table 4.6, was based on cost, annual energy yield for a given wind climate and here reliability considerations have been added using the approach described in Reference 15 of Chapter 5.

The evaluation showed that the indirect drive DFIG3IG, was the lightest, lowest cost solution using standard sub-assemblies, explaining why it is most widely used commercially. OEMs use generator and converter sub-assemblies close to industrial standards yielding standardisation, cost and reliability benefits. However, this system has wear in the gearbox and generator brush-gear and slip-rings and known unreliability in those areas. It also has a low energy yield due to the high gearbox losses. Since it uses a low cost, standard electrical machine and gearbox, future major improvements in performance or cost reduction cannot be expected.

The DDWRSGE appeared to be the heaviest, most expensive alternative and from Section 4.2 does not necessarily have the best reliability. The only commercially successful large direct drive WT OEM, Enercon, uses this configuration but they claim other benefits from it including immunity to problems from voltage disturbances due to grid faults, as a result of the use of a fully rated converter. But this sub-assembly is of particular concern having three times the number of parts to the DFIG3G partially rated converter, three times the cost and probably three times the failure rate [5]. The wind industry frequently misunderstands that the power converter is one of the highest cost, least reliable drive train items, not the gearbox, as frequently quoted. However, converter faults have low *MTTR*, unlike the gearbox, and it is also clear that substantial improvements are progressively taking place in power electronics, reducing cost and raising reliability.

In principle, the DDPMG should be the best solution because the generator only has one winding, does not have brushes or a gearbox but has a fully rated converter. An important attraction of this configuration is that the active generator material weight for the same air-gap diameter is nearly halved over the DDWRSGE, while the energy yield is a few percent higher giving the highest energy yield of the configurations considered. However, compared to indirect drive systems, it is more expensive. Further improvements of this configuration may be expected because of decreasing power electronics costs and further optimisation and integration of the generator system. However, the rising costs of permanent magnet materials are a current cause for concern.

The PMG1G, with a single-stage gearbox, is an interesting option because the electrical machine size is reduced by the higher speed and it is clear that this generator configuration with single-stage rather than three-stage gearboxes will be much more reliable. Polinder also made the case that this type of machine is used in other applications, for example ship propulsion, so development costs can be shared.

Surprisingly, the DFIG1G, with a single-stage gearbox, seems the most interesting choice in terms of energy yield divided by cost mainly because the lower converter rating results in a reduction of converter cost and losses. However, this system may be too specialised to attract electrical machine and gearbox OEMs and it is likely that the larger diameter, slower, less standard DFIG may suffer from unacceptable reliability.

Finally, an important aspect of any drive train configuration is the possibility of integral turbine, gearbox and generator design improving manufacture, transportation and installation, which may considerably affect the WT price.

4.6   Innovative WT configurations

Beyond the possibilities of current geared and direct drive wind turbine configurations, shown in Figure 4.2, there are a number of innovative new concepts now under consideration. These are summarised briefly in Figure 4.9, expanding the classification in Reference 1. They are described here being divided into electrical or hydraulic options as follows and all offer potential reliability benefits:

- Type C′, a derivative of Type C using an LV brushless doubly fed induction generator (BDFIG) instead of the DFIG, removing the need for brush-gear and slip-rings, which affect the maintenance of DFIGs but also offers a lower speed generator, allowing the use of a two-stage rather than three-stage gearbox.
- Type C″ another derivative of Type C using an LV WRIG with a self-driven three-phase AC brushless exciter, feeding the rotor through a two-quadrant excitation-rated converter, removing the need for brush-gear and slip-rings.
- Type D′, a derivative of Type D, the geared drive WT but using a single- or two-stage gearbox, low-speed generator and fully rated converter, thereby raising the gearbox reliability, gaining the power quality advantages of the power converter and eliminating brush-gear and slip-rings.
- Type E, a hydraulic arrangement based on a conventional geared drive train but using a limited speed range hydraulic torque converter to drive an MV WRSGE. The advantage of this arrangement is that a power electronic converter and transformer are eliminated by synchronous generation at MV, an example drive train is shown in Figure 4.10.
- Type E′, an innovative hydraulic solution using Digital Drive Technology (DDT) from Artemis Innovative Power with the turbine driving a slow speed hydraulic pump, which feeds a high-speed hydraulic motor with high-pressure hydraulic fluid. That motor drives an MV WRSGE. The advantage of this arrangement with synchronised MV generation is that gearbox, power electronic converter and transformer are eliminated; however, the DDT hydraulic scheme is new and untried.

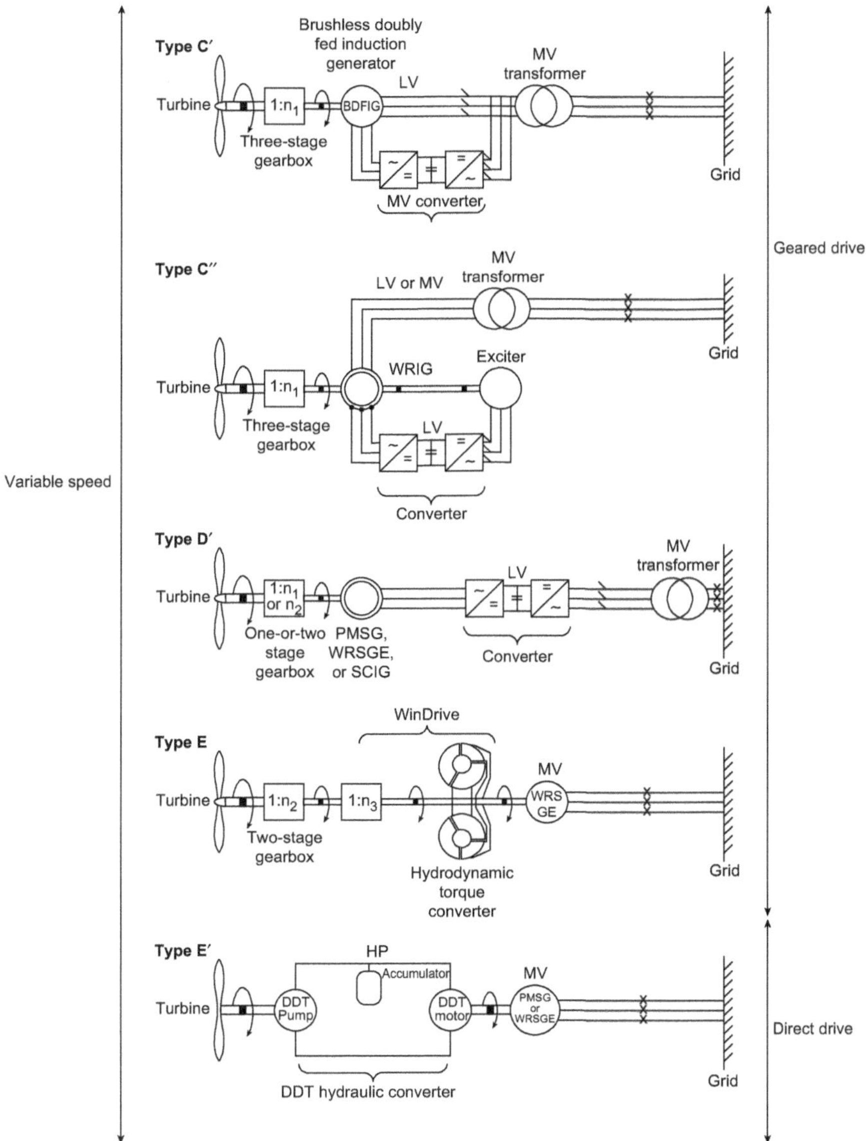

Figure 4.9   Summary of main electrical configurations for innovative WT drive trains (cf. Figure 4.2)

4.7 Summary

This chapter has shown that turbine configuration does have an effect on WT reliability but that there are some industry myths that are not supported by evidence. For example, it is simply not proven that a direct drive WT is more reliable

Figure 4.10   Example drive train Type E, an assembled DeWind D8.2 WT with Voith WinDrive hydraulic torque converter driving a 2 MW, 11 kV WRSGE [Source: Voith Drives]

than a geared drive machine. It is also clear that the influence of low reliability power electronics does not help the reliability case of fully rated converter WTs.

However, electrical sub-assemblies, such as the converter, appear to have lower *MTTRs* and improving reliability, whereas heavy mechanical sub-assemblies, such as the gearbox, have high *MTTRs* and are mature technology whose reliability is not improving. This suggests that in the long term all-electric WTs must have a more reliable future.

On the other hand, there are some emerging drive train technologies, such as semi-direct drives with single-stage gearboxes and low-speed LV permanent magnet generators, or hydraulic drive transmissions, which allow the use of fixed-speed MV generators, showing great promise for reducing weight, reducing the Balance of Plant (BOP) and improving reliability, but their full production capital costs are not yet known.

Recent experience has emphasised that any configuration can achieve reliability provided that the component sub-assemblies are well designed, manufactured, installed and maintained.

An important conclusion is that there is no clear ideal OWT configuration for reliability, rather that OEMs should ensure that drive train sub-assemblies are thoroughly tested before installation in the WT and that WT nacelles should be prototype load-tested or even production tested at load if they are to be installed offshore.

4.8   References

[1]   European Wind Energy Association, Wind Energy Factsheets, **10**, 2010.

[2]   Hansen A.D., Hansen L.H. 'Wind turbine concept market penetration over 10 years (1995–2004)'. *Wind Energy*. 2007;**10**(1):81–97

[3]   Polinder H., van der Pijl, F.F.A., de Vilder G.J., Tavner P.J. 'Comparison of direct-drive and geared generator concepts for wind turbines'. *IEEE Transactions on Energy Conversion.* 2006;**21**(3):725–33

[4]   Tavner P.J. 'Review of condition monitoring of rotating electrical machines'. *IET Proceedings Electrical Power Applications.* 2008;**2**(4): 215–47

[5]   Tavner P.J., Faulstich S., Hahn B., van Bussel G.J.W. 'Reliability & availability of wind turbine electrical & electronic components'. *European Journal of Power Electronics.* 2011;**20**(4):45–50

[6]   IEEE, Gold Book. *Recommended Practice for Design of Reliable Industrial and Commercial Power Systems.* Piscataway: IEEE Press; 1990

[7]   Alewine K., Chen W. 'Wind turbine generator failure modes analysis and occurrence'. *Proceedings of Windpower*, May 24–26, Dallas, Texas, 2010

[8]   O'Donnell P. 'IEEE reliability working group, report of large motor reliability survey of industrial and commercial installations. Parts I, II and III'. *IEEE Transactions on Industry Applications.* 1985;**21**:853–72

[9]   Albrecht P.F., McCoy R.M., Owen E.L., Sharma D.K. 'Assessment of the reliability of motors in utility applications-updated'. *IEEE Transactions on Energy Conversion.* 1986;**1**:39–46

[10]   Thorsen O.V., Dalva M. 'A survey of faults on induction motors in offshore oil industry, petrochemical industry, gas terminals and oil refineries'. *IEEE Transactions on Industry Applications.* 1995;**31**(5):1186–96

[11]   Spinato F., Tavner P.J., van Bussel G.J.W., Koutoulakos E. 'Reliability of wind turbine sub-assemblies'. *IET Proceedings Renewable Power Generation.* 1995;**3**(4):1–15

Chapter 5

Design and testing for wind turbine availability

5.1 Introduction

The high penetration of wind power into power systems will have several impacts on their planning and operation. One of these will be the effect on power system reliability, emphasised because wind power is intermittent, so WT reliability delivering this power is becoming an essential consideration. Due to the competitive environment, power generation industry developers and operators usually prefer the most economically productive WT configurations. This must take into consideration work like that shown in Section 5.2. However, through-life productivity must also be considered, emphasised in offshore operation, where access is difficult and otherwise productive WTs may be unproductive because of small but unresolved faults, see Reference 1.

Long-term cost analysis of WTs, including both first investment and operation and maintenance (O&M) costs, will result in better WT configuration choices, but this is only possible if such analysis includes the reliability of the different WT technologies.

Reliability of WTs as part of a larger power system has been assessed in a number of references [2–4] considering the wind as a stochastic process, using an appropriate time series to model the wind resource input combined with the power–speed curve of an appropriate WT.

There have been few studies of the reliability of WTs as isolated systems rather than as part of a large power system [5], although Xie and Billinton [6] do consider the impact of WT reliability in the overall reliability of the power system. This chapter follows the previous chapter focusing on the design of the WT, consisting of several mechanical, electrical and auxiliary assemblies, as part of a larger wind farm, showing the methods that can be applied to achieve the reliability objective.

Reliability analysis methods in the initial stages of power generation system design are usually qualitative, depending on comparison with data from similar systems, whereas after several years power generation reliability analysis can become more quantitative as valid field statistics data are generated.

5.2 Methods to improve reliability

5.2.1 Reliability results and future turbines

The results presented in Chapter 4 were all obtained on existing WTs of historic design of size ranging from 200 kW to 2 MW.

To what extent can these data be used to predict the reliability performance of new designs of WT of much larger size, say 3–10 MW?

Reliability analysis is of necessity backward looking and rarely produces data less than 5 years old; however, its advantage is that data are numerical and comparable. It is proposed that the WT failure rates shown in Figures 3.5 and 3.6 could be used as a datum against which future designs should be measured. For example, while an average failure rate of 1 failure(s)/turbine/year could be acceptable onshore, it is unlikely to be acceptable offshore where access may be limited to one visit a year.

The WT sub-assembly failure rates can also be used as a datum for comparison between different concepts and designs; however, the *MTTR* must also be considered, as the gearbox data have shown.

Reliability improvement analysis will be useful for WT and sub-assembly OEMs to define where design and testing effort should be deployed to improve future reliability.

5.2.2   Design

One simple approach to improve reliability, taken by Enercon and other WT OEMs, has been to remove the gearbox and use a direct drive configuration. Enercon also adopted an all-electric approach, avoiding the use of hydraulics for pitch or yaw control. Comparison between direct and geared drive WTs, raised by Polinder *et al.* [7], has shown [8] the following:

- From Figure 3.4, direct drive WTs do not necessarily have better reliability than geared drive WTs. In Figure 3.4, the direct drive E40 has a higher failure rate than its geared drive partners of the same size, whereas the direct drive E66 has a lower failure rate than its partners, although the E66 data is rather limited in the number of WTs.
- From Figure 4.3, the aggregate failure rates of generators and converters in direct drive WTs are generally greater than the aggregate failure rate of gearboxes, generators and converters in geared WTs. Therefore, the price paid by direct drive WTs for the reduction of failure rate by the elimination of the gearbox is a substantial increase in failure rate of electrical-related sub-assemblies.
- On the other hand, from Figure 3.5 it can be seen that the *MTTR* of electronic sub-assemblies is lower than the MTTR of gearboxes.
- From Figures 4.3(b) and (c), the failure intensities of larger direct drive generators are up to double that of the geared drive generators of similar size. The following explanation is offered. The direct drive machines in these machines were wound rotor synchronous generators with high pole pair number, incorporating a large number of rotor and stator coils, whereas the geared drive machines are four or six switchable pole, high-slip, induction generators or DFIGs, with far fewer coils. It is suggested that the disparity in failure intensities is because of the following:
  - ○ The much larger number of coils in the direct drive machine. The failure rate could be improved by replacing field coils by permanent magnets, but this would introduce other, reactive control issues.

    o   The larger diameter of the direct drive generator, making it difficult to seal the more numerous coils from the environment, exposing coil insulation to damage because of the air contaminants and environmental humidity.

    o   Insufficient standardisation in the manufacture of the large direct drive machines, as a consequence of smaller production runs, compared to the more common DFIG. From a general consideration of direct drive or geared concept WTs, the following issues arise associated with the design:

- The reliability of these WT generators, from Figure 4.4, is worse during early operational life than that achieved by generators in other industries.

- From Figure 4.5, the reliability of these WT gearboxes are seen to be that of a mature technology, constant or slightly deteriorating with time. The reliabilities are comparable with those obtained by gearboxes in other industries. Therefore, substantial improvements in the designed reliability of these gearboxes are unlikely in the future, although design improvements in gearboxes for newer, larger designs of WTs are being actively pursued and it appears that maybe a greater onus is being placed on WT gearbox reliability by the stochastically varying torque to which it is subjected.

- From Figure 4.6, the reliability of these WT converters is considerably worse throughout their operation than achieved by converters in other industries.

- The *MTTR* of electrical components is relatively low and industrial experience suggests that electrical sub-assemblies are more amenable to reliability improvement than mechanical sub-assemblies, for example the gearbox. Therefore, an all-electric direct drive WT may ultimately have an intrinsically higher availability than a geared drive WT.

- From the observations, above improvements in generator and converter reliability design will be crucial to improving the reliability of both direct drive and geared concept WTs and this design information is exceptionally important for OWTs.

This chapter will go on to show that there is more that can be done to promote reliability during the design stage, Section 5.3, than simply changing the overall WT configuration or concept.

5.2.3 Testing

Testing of sub-assemblies, particularly converters and generators, can encourage the achievement of higher WT reliability at the start of operational life by eliminating early failures. A suggestion is that offshore WTs nacelles could be tested complete, at full or varying load, at elevated temperature, to accelerate the occurrence of early failures. This is a standard practice in the electrical machine and gearbox industry where prolonged heat runs at elevated temperatures are done as type tests on new products. These type tests are then repeated on individual machines from batch sizes specified, for example, by IEC Standards 60034 and

61852. It is also a standard practice, in the volume production of low-rating power converters, <100 kW, to routinely age key converter sub-assemblies and then carry out extended load tests on assembled converters from batch sizes specified, for example, by IEC Standard 60700, to identify generic weaknesses before despatch.

The issues of testing are discussed further in this chapter in Section 5.4.

5.2.4 Monitoring and O&M

The improving reliability of generators and converters in Figures 4.4 and 4.6 indicates that O&M activities are already having a reliability effect. Condition monitoring measures machine performance indicating the need for remedial action when performance deteriorates. The wind industry has applied SCADA and CMS systems to WTs and most wind farms now have a SCADA system providing data to remote control rooms. However, agreement has not yet been reached on processing the large quantities of data generated to indicate incipient failures. O&M methods need to use this information to predict failure and thereby schedule maintenance, although work is currently going on in this area of O&M [9]. If the design and testing suggestions above are developed and the monitoring techniques are resolved, the O&M approach will require

- maintenance based on the measured condition of the WT so that failures of vital sub-assemblies like the generator, gearbox and converter can be pre-empted;
- the provision of adequate spares to reduce downtime when maintenance on the basis of condition takes place.

These issues will be raised in the following sections of this chapter and in Chapters 6 and 8.

5.3 Design techniques

5.3.1 Wind turbine design concepts

WT OEMs moving into the offshore market are concerned to develop designs appropriate to that market. Some OEMs have deployed offshore WTs designed for onshore and this has given rise to a number of operational problems, for example associated with particular maintenance operations, implicit in their design, which are untenable offshore.

Therefore, OEMs have been anxious to develop new or modified WT designs, appropriate to the more onerous offshore environment. Some WT OEMs and their investors have favoured particular design concepts for this application, for example direct drive as opposed to geared drive, or hydraulic transmission as opposed to electrical conversion, to avoid perceived onshore WT reliability problems. The issue of direct vs geared drive has been investigated in Reference 7, their reliability in Reference 8 and their electrical parts in Reference 1. The issue of hydraulic transmission vs electrical conversion has not been investigated from a reliability point of view, although it could be using the methods described in Reference 8.

As a general principle, it would be unwise to introduce offshore a new design concept that had not been thoroughly pre-tested and been exposed to onshore operation. Furthermore, it is a mistake for the wind industry to imagine that any particular WT design concept, whether direct or geared drive, all-electric, hydraulic or mechanical transmission, is likely to be a panacea for offshore operation.

It is clear that there are good examples of all those technologies working reliably onshore, which can be made to succeed offshore but only with adequate pre-testing and an appropriate O&M regime enforced in the offshore environment.

As concluded in the last chapter, experience has emphasised that any concept can achieve reliability provided that the component sub-assemblies are well designed, manufactured, installed and maintained. The object here is to highlight precautions that can be taken during design to raise reliability.

5.3.2   Wind farm design and configuration

Reliability depends not only on the WT but also on the design of the wind farm in which the WT is situated and its configuration, which contains not only the WT but also collector cable arrays, substations, cable connection to shore and a shore substation [10]. Figure 5.1 shows a typical radial cable configuration for a large wind farm.

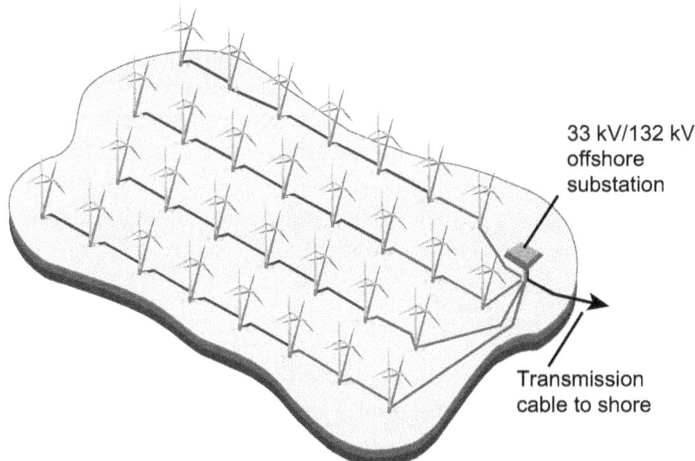

Figure 5.1   Radial configuration of an offshore wind farm with 33 kV collector voltage and 132 kV grid connection

As yet there has been no published FMEA for a wind farm array but the key issues to consider are as follows:

- Individual WT transformer and switchgear arrangements for connection to the 20–33 kV cable collector array

- Configuration of the 20–33 kV cable collector array itself, including any switchgear to allow sub-group isolation within that array
- Configuration of the collector substation including the collector array electrical protection
- Switchgear, transformers for the cable connection to shore
- Shore substation and protection for the cable connection to shore

An important issue here is the degree of redundancy incorporated into the collector cable array and offshore substation. Early offshore wind farms had radial collector arrays, as in Figure 4.1, meaning that a single fault in a radial spur would interrupt power flow from the whole spur. However, by introducing some ring capability, with additional cable routes and switchgear, there can be an improvement in overall wind farm reliability and availability, by providing alternative power flow routes in the event of a failure in the collector network, but this adds to the project cost. Cable arrays have been investigated in Reference 7.

5.3.3   Design review

A procedural method for raising prospective offshore wind farm and OWT reliability is to apply Design Review procedures in the development phase. A process for OWT design, recommended by the draft standard, is shown in Figure 5.2.

OWTs will be qualified for the rated wind speed and wind class (see Table 5.1) with a design lifetime of at least 20 years for wind turbine classes I to III.

In Table 5.1, the parameter values apply at hub height and V_{ref} is the reference wind speed average over 10 minutes. A designates the WT category for higher turbulence characteristics, B for medium turbulence characteristics, C for lower turbulence characteristics and I_{ref} is the expected value of the turbulence intensity at 15 m/s.

An important issue here is that WT reliability, described briefly at the start of Chapter 2, can be considered to consist of

- structural reliability;
- electro-mechanical reliability;
- control system reliability.

The process of certified design in Figure 5.2 is mainly directed towards structural survivability when the OWT is subjected to the extreme events during its planned life.

Such analysis is made more complex because the turbine is also subject to aleatory uncertainty due to the stochastic effects of the weather itself, the wind from which the machine extracts energy and, in the case of OWTs, the combined effects of wind and waves on the structure and of corrosion.

The impact of these extreme events is of primary importance to the OWTs' structure, vital to its survival, but does not impact upon the day-to-day operation and normal life, which depend upon the electro-mechanical and control reliability.

The reliability aspects of design must therefore concentrate upon these electro-mechanical and control issues and be wrapped around the process of Figure 5.2 as

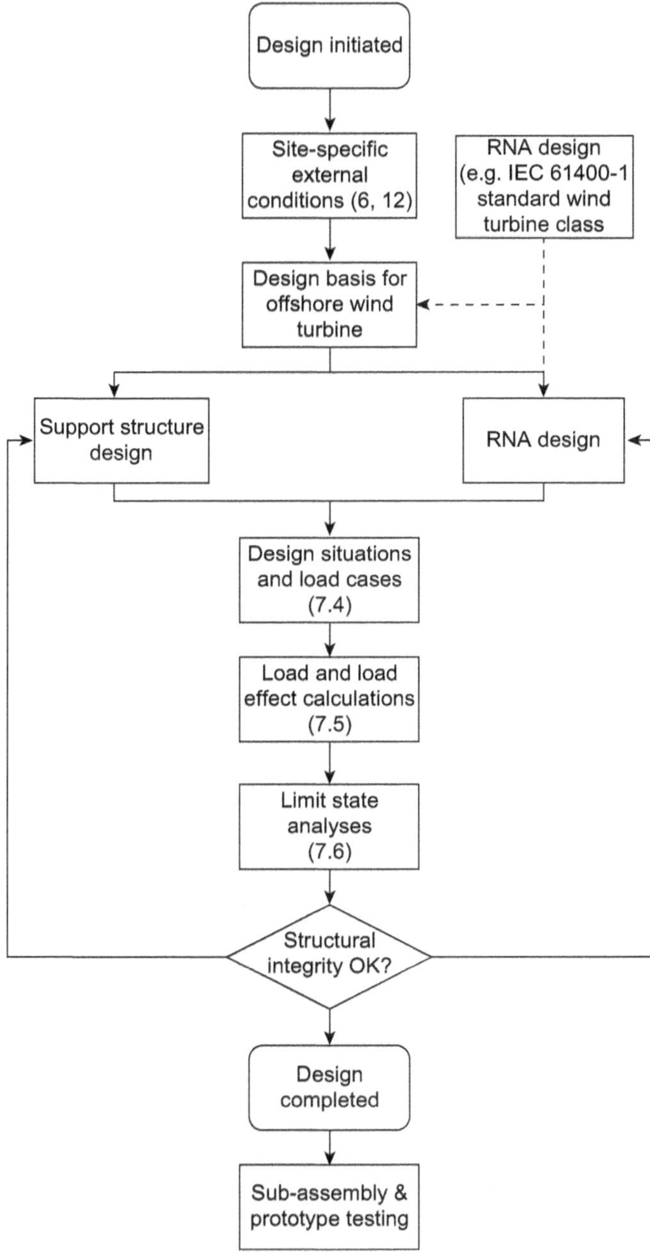

Figure 5.2   Description of the design process for an offshore WT [Source: [11]]

part of the Design Review process and it will be essential that the process is based upon genuine reliability data, either obtained from earlier developments of the OWT or from surrogate data from the offshore industry or from WTs of similar design, such as that described in Chapter 3.

Table 5.1   Basic parameters for wind turbine classes

Wind turbine class	I	II	III	S
V_{ref} (m/s)	50	42.5	37.5	Values specified by the designer
$A\ I_{ref}$ (−)	0.16			
$B\ I_{ref}$ (−)	0.14			
$C\ I_{ref}$ (−)	0.12			

[Source: [12]]

It would also be advantageous to combine the Design Review process with an FMEA/FMECA process, such as that described in Section 5.3.4.

5.3.4   FMEA and FMECA

Failure Modes and Effects Analysis (FMEA) or Failure Modes and Effects and Criticality Analysis (FMECA), where failure rates are considered, are the best candidates for design stage reliability analysis as part of RMP. The process is well defined [13] and has been used for many power generation engineering systems, although perhaps less with an emphasis on availability than concern for safety, design assurance or the avoidance of specific observed in-service failure modes.

The FMEA is a powerful design tool that provides a means, from a risk point of view, of comparing alternative machine configurations; it is also useful for considering designs improvements for a technology that is changing or increasing in rating, as WT configurations are.

The FMEA is a formalised but subjective analysis for the systematic identification of possible root causes and failure modes and the estimation of their relative risks.

The main goal is to identify and then limit or avoid risk within a design. Hence, the FMEA drives towards higher reliability, higher quality and enhanced safety.

Since FMEA is used by various industries, including automotive, aeronautical, military, nuclear and electro-technical, specific standards have been developed for its application. A typical standard will outline Severity, Occurrence and Detection rating scales as well as examples of an FMEA spreadsheet layout. Also, a glossary will be included that defines all the terms used in the FMEA. The rating scales and the layout of the data can differ between standards, but the processes and definitions remain similar, for example:

- SAE J 1739 was developed as an automotive design tool and Ford has used it as a Design Review process.
- SMC Regulation 800-31 was developed for aerospace.
- IEC 60812:2006 [13] is a general standard.
- MIL-STD-1629A (1980) [14] drafted by the US Department of Defense is the most widely used FMECA standard with over 30 years development and usage, having been employed in many different industries for general failure analysis. Due to the complexity and criticality of military systems, it provides a reliable foundation on which to perform FMEAs on a variety of systems. It also

contains formulae for predicting the failure rates of electrical and electronic systems, whose coefficients are based on accelerated life tests.

It can also be used to assess and optimise maintenance plans. An FMEA is usually carried out by a team consisting of design and maintenance personnel whose experience includes all the factors to be considered in the analysis. The causes of failure are root causes, and may be defined as mechanisms that lead to the occurrence of a failure. While the term failure has been defined, it does not describe the mechanism by which the component has failed. Failure modes are the different ways in which a component may fail. It is vitally important to realise that a failure mode is not the root cause of failure, but the way in which a failure has occurred. The effects of one failure can frequently be linked to the root causes of another failure.

The FMEA procedure assigns a numerical value to each risk associated with causing a failure, using Severity, Occurrence and Detection as metrics. As the risk increases, the values of the risk rise. These are then combined in a Risk Priority Number (RPN), which can be used to analyse the system, where RPN is calculated by multiplying the Severity, Occurrence and Detection of the risk:

$$RPN = Occurrence \times Severity \times Detectability \tag{5.1}$$

By targeting high RPN values, the most risky elements of a design can be addressed.

Severity refers to the magnitude of the end effect of a system failure mode. The more severe the consequence, the higher the value of severity will be assigned to the effect.

Occurrence refers to the frequency that a root cause is likely to occur, described in a qualitative way, that is, not in the form of a period of time but rather in terms such as remote or occasional.

Detection refers to the likelihood of detecting a root cause before a failure can occur.

In conventional FMEA, the Severity, Occurrence and Detection factors are individually rated using a numerical scale, typically ranging from 1 to 10. These scales, however, can vary in range depending on the FMEA standard being applied. However, for all standards, a high value represents a poor score, for example catastrophically severe, very regular occurrence or impossible to detect. Once a standard is selected it must be used throughout the FMEA. In this section, Reference 13 will be used but with some amendment, principally to change the Severity, Occurrence and Detection criteria by which the RPN is calculated. These modifications were necessary to make the FMEA methodology more appropriate to WT systems.

The modified Severity scale and criteria are shown in Table 5.2. The scale of 1–4 in Reference 13 was maintained but changes were made to the category criteria definitions to emphasise their implications for a WT.

An Occurrence scale and criteria modified from Reference 13 are tabulated in Table 5.3. Arabian-Hoseynabadi *et al.* [15] have shown that Severity can be related to $1/\mu = MTTR$.

Finally, the number of Detection levels were reduced, according to Reference 15, to 2 as shown in the modified Detection scale and criteria tabulated in

Table 5.2   Severity rating scale for a WT FMEA

Scale no.	Description	Criteria
1	Category IV (minor)	Electricity can be generated but urgent repair is required
2	Category III (marginal)	Reduction in ability to generate electricity
3	Category II (critical)	Loss of ability to generate electricity
4	Category I (catastrophic)	Major damage to the turbine as a capital installation

Table 5.3   Occurrence rating scale for a WT FMEA

Scale no.	Description	Criteria
1	Level E (extremely unlikely)	A single failure mode probability of occurrence is less than 0.001
2	Level D (remote)	A single failure mode probability of occurrence is more than 0.001 but less than 0.01
3	Level C (occasional)	A single failure mode probability of occurrence is more than 0.01 but less than 0.10
4	Level A (frequent)	A single failure mode probability greater than 0.10

Table 5.4   Detection scale for a WT FMEA

Scale no.	Description	Criteria
1	Almost certain	Current monitoring methods almost always will detect the failure
2	Almost impossible	No known monitoring methods available to detect the failure

Table 5.4. Arabian-Hoseynabadi *et al.* [15] has shown that Occurrence can be related to $\lambda = 1/MTBF$.

It can be concluded from Tables 5.2 to 5.4 that with these gradations the minimum RPN for any root cause is 1 and the maximum is 32. As long as the rating scales of a selected FMEA procedure remain fixed between alternative WT designs, they can be used for the comparison of those alternatives and identification of critical assemblies. Defining these three criteria tables based on MIL-STD-1629A standard [14] is the first step in performing an FMEA. As mentioned before, the basic principles of an FMEA using different standards are similar and simple:

- The system to be studied must then be broken down into its sub-systems, assemblies, sub-assemblies and components.
- Then, for each sub-system, assembly, sub-assembly and component all possible failure modes must be determined.
- The root causes of each failure mode must be determined for each sub-system, assembly, sub-assembly and component.

- The end effects of each failure mode must be assigned a level of Severity, and every root cause must be assigned a level of Occurrence and Detection.
- Levels of Severity, Occurrence and Detection are multiplied to produce the RPN.

Therefore, the first stage in the FMEA procedure is obtaining a comprehensive understanding of the WT system and its main assemblies. This is set out in Appendix 2 of this book based upon the experiences of the ReliaWind Consortium.

The FMECA will require the designer to define failure modes and root causes for each sub-assembly in the wind turbine. Experience has shown that individual designers can generate a very wide spread of idiosyncratic failure modes and root causes, depending on their individual expertise and knowledge of the WTs field operation. The author's experience suggests that it makes the FMECA more meaningful if generic failure modes and root causes are adopted, at least initially, and that these form a standard for the designers to use across the sub-assemblies. A list of generic failure modes and root causes is shown in Table 5.5, which has been used in the author's paper and can be the basis for future development in specific FMECAs.

Table 5.5   Suggested generic failure modes and root causes for a WT FMEA

Failure modes	Failure root causes
Structural failure	Design defect
Electrical failure	Material defect
Mechanical failure	Installation defect
Software or control failure	Maintenance defect
Insulation failure	Software defect
Thermal failure	Corrosion
Mechanical seizure	Misalignment
Bearing failure	Low-cycle fatigue
Component fracture or material failure	High-cycle fatigue
Seal failure	Mechanical wear
Contamination	Lack of lubrication
Blockage	Thermal overload
	Electrical overload
	Weather incident
	Grid incident

Software can be used to facilitate the FMECA and other system reliability studies. The author has had experience of the following software packages:

- ReliaSoft, XFMEA [20]
- Isograph, Reliability Workbench [21]
- PTC-Relex, Reliability Studio 2007 V2 [22]

Users will need to evaluate these packages individually for their own needs. The more sophisticated aspects of the packages allow various forms of reliability modelling to be used, allowing access to database reliability information and discipline for the analysis structure. However, for an FMEA, it is perfectly possible to assemble a professional analysis on the basis described above solely using an Excel spreadsheet.

There has been one published account of an FMECA applied to a WT [15], and an EU FP7 project [16] proposed the application to individual WT and WT sub-assembly OEMs, with preliminary results reported in Reference 17, a full report in Reference 18 and a detailed application to a common WT type with three different drive trains in Reference 19. Other relevant reliability studies on drive trains and the electrical sub-assemblies of them are given in References 23 and 24.

A useful analysis from the FMECA results is the occurrence frequency of the different failure modes and root causes. The repetition rate of these limited numbers of failure modes and root causes can be analysed for the WT being considered and this gives a good ranking for the key root causes to be mitigated and failure modes to be detected. Counting these failure modes and root causes over the whole FMECA can give histograms for each, this was done in Reference 15, identifying the top 10 failure modes and root causes in Figure 5.3.

As can be seen from Figure 5.3, the most significant failure mode was material failure, so improved material quality in WTs must be key point for reliability enhancement. It is worth mentioning that these failure mode frequencies are based upon FMEA results and not on chronological data of wind turbine performance. Similarly, the most frequent root cause is corrosion, which affects the material quality. This will be more important in future offshore WTs, so remedial design actions in this regard must be considered.

Identifying the most frequent failure modes and root causes will assist design improvement and maintenance optimisation. A cost–benefit analysis for reducing WT failures could be conducted based on a priority list of the most frequent failure modes. A similar analysis could also be considered based on failure modes severity, for example, by summating the severity of each failure mode in the FMEA, ranking the results and considering the costs incurred to alter the ranking.

5.3.5   *Integrating design techniques*

The author's proposed method of integrating these above design techniques during design, pre-production and production tests for an OWT is shown in Figure 5.4.

This is based upon the construction of a pre-production prototype OWT, development of an integrated SCADA/CMS system for the OWT, construction of a production prototype and construction of production machines. The design process needs to be integrated by a process of testing, data collection and checking. In this case the FMECA document is used as a means to check progress.

This process would need to be extended after the design and prototype building phase to include commissioning and operations as shown in the next chapter.

5.4   Testing techniques

5.4.1   *Introduction*

Section 5.2.3 emphasised the importance of testing as a further means to raise WT reliability. All testing is intended to raise the reliability of components by lowering

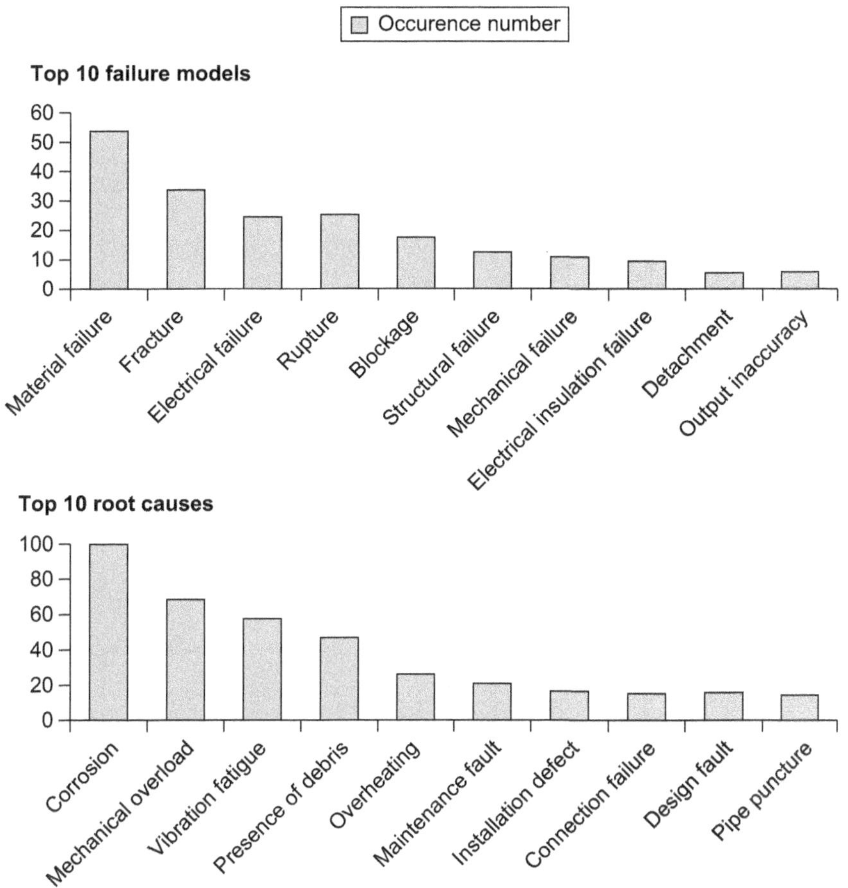

Figure 5.3   Top 10 failure modes and root causes from the FMEA in Reference 15, using generic examples as in Table 5.5

failure rate, λ, or increasing *MTBF*, see the bathtub curve (Figure 2.4), repeated here in Figure 5.5(a).

The effects of pre-production testing can be seen in Figure 5.5(b). However, this testing can be broken down into a number of different activities at different stages of the OWT design, as described in the following sections.

5.4.2   Accelerated life testing

Accelerated life testing (ALT) is aimed at measuring component and sub-assembly failure rates in a controlled test environment, whose conditions can be varied in such a way that the ageing process is accelerated. The acceleration of the test is achieved by applying a stress that is greater than that encountered in service but not

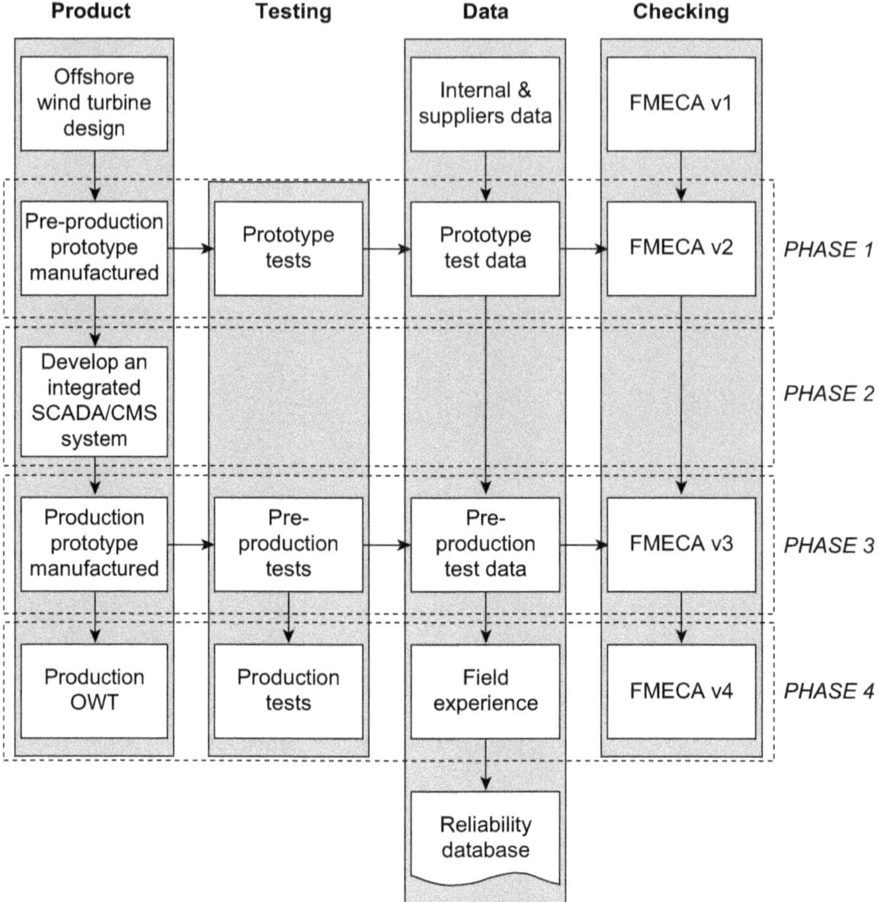

Figure 5.4   Proposal for using FMEA as an OWT Design Review tool during design and manufacture

beyond technological limits. This shortens the time to failure but without altering the failure mechanism, which is assumed to be activated selectively by the increased stress producing an acceleration factor, *Acc* [25].

ALT aims to collect reliability data for individual components or sub-assemblies to be used in reliability analysis to lower the intrinsic failure rate, λ, of a whole system, along the lines shown in Figure 5.5(b). ALT was originally developed for electronic components, where many of the ageing processes are driven by temperature, so acceleration is achieved simply by raising the temperature and evaluating *Acc* using Arrhenius Rate Law, but has been expanded for use with electro-mechanical sub-assemblies.

The object is to derive detailed life reliability curves for individual components or sub-assemblies, such as those shown in Figure 2.5, in the environmental conditions they are likely to encounter in service. For OWTs, this should cover:

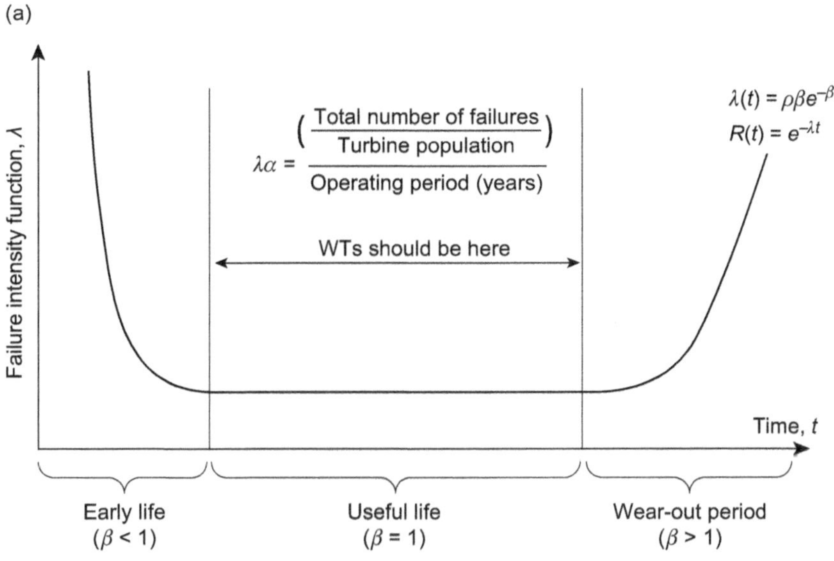

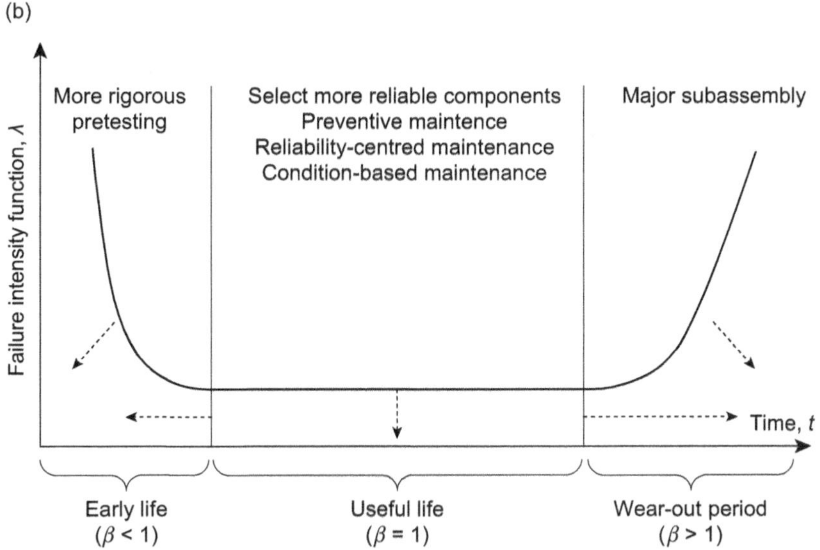

Figure 5.5   Bathtub curve of failure intensity showing effect of testing. (a) Bathtub curve of failure intensity; (b) effects on the curve of testing and maintenance

- High or low temperatures
- High humidity
- Saliferous atmosphere

ALT can provide core data for reasonable component and sub-assembly reliability predictions to be used in an FMECA for a prospective OWT design.

Without ALT, designers need to obtain data from free or commercial databases, sometimes available from WT OEM sub-suppliers. Release of such data can be part of the sub-suppliers procurement contracts.

5.4.3   Sub-assembly testing

In the absence of ALT data, or the ability to do ALT testing, OWT assemblies and sub-assemblies need to be thoroughly pre-tested in a low-cost, benign test bed environment, perhaps at elevated load or temperature, to secure more reliable offshore deployment.

This will reduce the early life failures at the start of the bathtub (Figure 5.4(b)), reducing early failures in service. That process must start with the sub-assemblies most at risk, identified from public domain data (Figures 3.5 and 3.6), or data available to the OWT OEM and its sub-suppliers from operational experience with previous models. These sub-assemblies may be

- mechanical, such as pitch motion units, lubricating oil systems or hydraulic power-packs; or
- power electronic modules, such as generator- and grid-side inverters; or even
- controller sub-assemblies, such as the yaw, pitch and generator controllers.

The issue of sub-assembly testing has been shown in other industries, for example in power electronic variable speed drives, to be of particular importance for highly complex electrical and electronic sub-assemblies with high failure rates and low *MTTR*, with the potential for great reliability improvement. Considerable efforts have been made in the electronics industry to improve sub-assembly reliability through systematic testing and supplier quality control, see Reference 25.

Another benefit of controlled sub-assembly testing is that it generates the numerical data, which, when added to that produced from ALT, progressively builds an objective reliability model for a prototype WT and provides the basis for future procurement quality control.

5.4.4   Prototype and drive train testing

Despite the accumulation of data from ALT and sub-assembly testing, there will still be a need to prototype test the OWT itself or at the least major sub-systems. Chapter 4 has shown that drive train reliability is a major cause for concern, less because of its failure rate than for the excessive *MTTR* and consequent drive train failure costs.

This is therefore becoming a major development area for OWT OEMs and their drive train sub-assembly suppliers who are conducting a number of such tests, see, for example, Figure 5.6 showing a 2.5 MW drive train under test, exemplifying this trend.

This process is particularly important for offshore operations, where high offshore installation and access costs must encourage WT OEMs and developers to reduce subsequent interventions.

Figure 5.6   Drive train test rig mounting a Samsung 2.5 MW drive train at the National Wind Technology Center [Source: National Renewable Energy Laboratory, USA]

Figure 5.7   Planned 15 MW drive train test rig [Source: National Renewable Energy Centre, UK]

Figure 5.7 shows an example of the world's largest planned drive train test rig, valued at >£30 million, which is intended to apply the torque and force components expected on modern large OWTs to prototype drive trains. Again, a major motivation for pursuing this kind of test exercise, which will be costly, is to

- test novel arrangements under known offshore torque and lateral force transients;
- accumulate test information to inform drive train sub-assembly testing;
- involve gearbox, generator, converter or hydraulic sub-assembly OEMs in the development of a robust drive train;
- de-risk new drive train concepts.

5.4.5   *Offshore environmental testing*

Part of the offshore situation is ensuring the reliability of parts to exposure to the more difficult ambient environment, from the point of view of temperature, humidity and saliferous atmosphere. This will be mitigated in most new OWT designs by nacelle sealing and the use of pressurised air treatment units. But part of the testing process must include exposure to those conditions. This can be achieved most cheaply at the sub-assembly stage, even if it does not contain the detailed ALT testing referred to in Section 5.4.2.

However, exposure of combined systems in some pre-production offshore test sites will generate a degree of experience and data to control the procurement of those sub-assemblies.

However, the offshore oil and gas industry has shown that an important factor in achieving reliability in the harsh offshore is by ensuring that the interfaces between pre-tested sub-assemblies; wiring, pipe work and junction boxes, are of the highest physical quality using stainless steel enclosures and high-quality pipework and armoured wiring as shown in Figure 5.8.

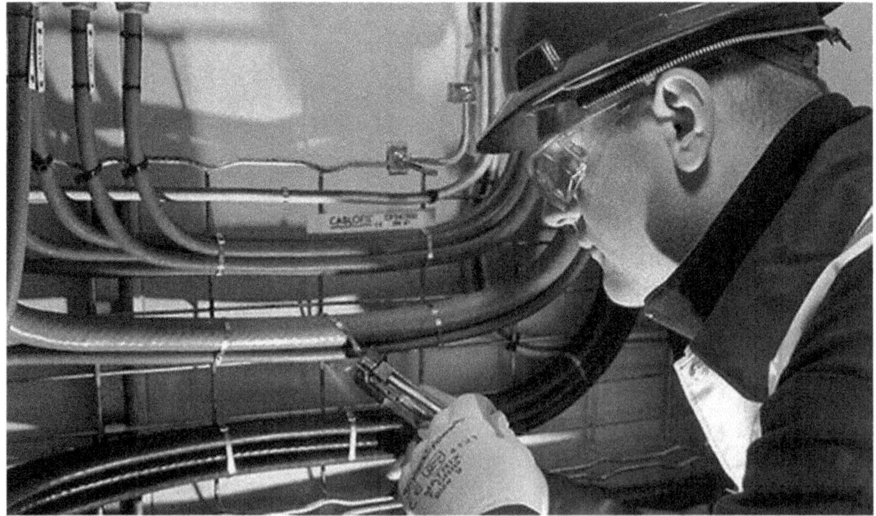

Figure 5.8   High-quality offshore wiring and cabling [Source: Cablofil]

5.4.6   Production testing

Considerable attention has been focused on gearbox reliability, and some gearbox OEMs are routinely back-to-back production testing their products as shown in Figure 5.9.

Figure 5.9   Back-to-back testing of two 3 MW wind turbine gearboxes [Source: Hansen Transmissions]

However, Chapter 4 has shown that the converter too is a high-risk sub-assembly, even if its *MTTR* is lower, and Figure 5.10 shows the routine production testing of a large converter prior to despatch.

An important innovation for OWTs may be the back-to-back testing of complete nacelles at variable and full power before despatch from the factory, as was suggested in Reference 8.

However, OWT OEMs will need to devise efficient means to achieve these processes cost-effectively in a timely way.

5.4.7   Commissioning

Once an OWT has been installed, testing is not complete until commissioning testing is finished (Figure 5.11). Commissioning testing has an important further part to play in identifying early failures and resolving them early in operational life. High-quality commissioning also plays a major part in the accurate setting of alarms of SCADA and CMS systems, which is crucial for the reliable operation of the OWT but requires considerably more resource to execute effectively in the offshore environment.

Figure 5.10   Production testing of a large wind turbine converter [Source: ABB Drives]

5.5   From high reliability to high availability

5.5.1   *Relation of reliability to availability*

The relationship between reliability and availability is shown in Figure 1.10 and the relationship of reliability and availability to cost of energy is shown in Figure 1.14. The processes described in this chapter are designed to ensure the prospective availability of an OWT in a wind farm. But these processes cannot deliver high

Figure 5.11   Offshore wind turbine commissioning [Source: ABB Drives]

wind farm availability without additional support to maintain reliability in service. High availability in service depends upon installing a high-reliability OWT, as described above, and then on

- the offshore environment itself including access to the asset;
- the ability to detect and interpret low reliability in service;
- planned preventative and corrective maintenance in response to that detection and interpretation;
- a programme of asset management based on the above to consider the through-life performance of the asset.

5.5.2   Offshore environment

The environment plays an enormous part in our achievement of good performance. Offshore wind resource is strong but can also adversely affect performance because gusts and turbulence can damage the WTs and higher wind speeds lead to raised wave height limiting access.

5.5.3   Detection and interpretation

An OWT is a remote unmanned robotic power generation unit. Good availability performance cannot be achieved unless we can remotely and accurately detect performance deterioration and interpret it prior to action. Therefore, the installation of reliable and accurate SCADA and CMS systems will be essential to achieve this part of the offshore mission. It will be vitally important that the data arising from detection and interpretation is fed back into the offshore wind farm management system.

5.5.4   Preventive and corrective maintenance

The action needed from operation and detection of performance deterioration is an organised programme of maintenance.

This must include operational expense (OPEX) actions for preventive maintenance, based upon the OWT design, and corrective maintenance, driven by SCADA and CMS detection.

The results of maintenance must also be fed back into the database of reliability information for the offshore wind farm.

5.5.5   Asset management through life

Finally, the whole wind farm asset will need to be managed holistically against the energy produced, not only to justify the ongoing OPEX needed to maintain performance but also to plan for the large-scale capital expenditure (CAPEX) to maintain the asset over its planned life, including the longer-term deterioration and planned replacement of larger sub-assemblies, such as blades, gearboxes and generators.

5.6   Summary

This chapter has described techniques for improving the reliability of wind turbines during the design and manufacturing processes including design review, FMEA/ FMECA and testing. It has shown how these can be coupled together during the prototype process through testing of sub-assemblies and prototype turbines leading to the development of an RCM plan for full deployment of the product. The key to this is the availability and the generation of reliability data from these processes in a database including

- prior design field reliability data;
- accelerated lifetime testing;
- sub-assembly suppliers data;
- prototype test data;
- pre-production test data;
- commissioning test data;
- maintenance logs;
- SCADA/CMS in-service data.

Finally, this chapter has shown the connection between design for reliability and pre-testing and real operational availability, demonstrating what is needed to deliver low cost of energy from operational offshore wind farms. Chapter 6 will demonstrate our early experience with offshore wind farms, and later chapters will address individually the points raised in Section 5.5 to put these lessons to work to improve our performance in the field.

5.7   References

[1]   Faulstich S., Hahn B., Tavner P.J. 'Wind turbine downtime and its importance for offshore deployment'. *Wind Energy*. 2011;**14**(3):327–37

[2]   Wangdee W., Billinton R. 'Reliability assessment of bulk electric systems containing large wind farms'. *International Journal of Electric Power Energy System*. 2011;**29**(10):759–66

[3]   Karaki S.H., Chedid R.B., Ramadan R. 'Probabilistic performance assessment of wind energy conversion systems'. *IEEE Transactions on Energy Conversion*. 1999;**4**(2):212–17

[4]   Mohammed H., Nwankpa C.O.'Stochastic analysis and simulation of grid-connected wind energy conversion system'. *IEEE Transactions on Energy Conversion*, 2000;**15**(1):85–90

[5]   Karaki R., Billinton R. 'Cost-effective wind energy utilization for reliable power supply'. *IEEE Transactions on Energy Conversion*. 2004;**19**(2): 435–40

[6]   Xie K., Billinton R. 'Determination of the optimum capacity and type of wind turbine generators in a power system considering reliability and cost'. *IEEE Transactions on Energy Conversion*. 2011;**26**(1):227–34

[7]   Polinder H., van der Pijl F.F.A., de Vilder G.J., Tavner P.J. 'Comparison of direct-drive and geared generator concepts for wind turbines'. *IEEE Transactions on Energy Conversion*. 2006;**21**(3):725–33

[8]   Spinato F., Tavner P.J., van Bussel G.J.W., Koutoulakos E. 'Reliability of wind turbine sub-assemblies'. *IET Proceedings Renewable Power Generation*. 2009;**3**(4):1–15

[9]   Caselitz P., Giebhardt J. 'Fault prediction for offshore wind farm maintenance and repair strategies'. *Proceedings of European Wind Energy Conference, EWEC2003*. Madrid, Spain: European Wind Energy Association; 2003

[10]   Lundberg S. *Wind Farm Configuration and Energy Efficiency Studies-Series DC Versus AC Layouts*. PhD Thesis, Chalmers University, Sweden; 2006

[11]   IEC 61400-3:2010 Draft. Wind turbines – design requirements for offshore wind turbines. International Electrotechnical Commission

[12]   IEC 61400-1:2005. Wind turbines – design requirements. International Electrotechnical Commission

[13]   IEC 60812:2006. Analysis techniques for system reliability – procedure for failure mode and effects analysis (FMEA). International Electrotechnical Commission

[14]   MIL-STD-1629: Military Standard Procedures for Performing a Failure Mode, Effects and Criticality Analysis. US Department of Defense, 1980.

[15]   Arabian-Hoseynabadi H., Oraee H., Tavner P.J. 'Failure modes and effects analysis (FMEA) for wind turbines'. *International Journal of Electrical Power and Energy Systems*. 2010;**32**(7):817–24

[16]   ReliaWind. Available from http://www.reliawind.eu. [Last accessed 8 February 2010]

[17] Wilkinson M.R., Hendriks B., Spinato F., Gomez E., Bulacio H., Roca J., *et al.* 'Methodology and results of the ReliaWind reliability field study'. *Proceedings of European Wind Energy Conference, EWEC2010*; Warsaw: European Wind Energy Association; 2010

[18] ReliaWind, Deliverable D.2.0.4a-Report. *Whole system reliability model.* Available from http://www.reliawind.eu

[19] Tavner P.J., Higgins A., Arabian-Hoseynabadi H., Long H., Feng Y. 'Using an FMEA method to compare prospective wind turbine design reliabilities'. *Proceedings of European Wind Energy Conference, EWEC2010*; Warsaw: European Wind Energy Association; 2010

[20] Reliasoft Corporation. Available from http://www.reliasoft.com

[21] Isograph, FMEA Software. *Reliability Workbench 10.1.* Available from http://www.isograph-software.com

[22] Relex. *Reliability Studio 2007 V2.* Available from http://www.ptc.com/products/relex/

[23] Arabian-Hoseynabadi H., Tavner P.J., Oraee H. 'Reliability comparison of direct-drive and geared drive wind turbine concepts'. *Wind Energy.* 2010;**13**:62–73

[24] Arabian-Hoseynabadi H., Oraee H., Tavner P.J., 'Wind turbine productivity considering electrical sub-assembly reliability'. *Renewable Energy.* 2010;**35**: 190–97

[25] Birolini A. *Reliability Engineering, Theory and Practice.* New York: Springer; 2007. ISBN 978-3-538-49388-4

Chapter 6

Effect of reliability on offshore availability

6.1 Early European offshore wind farm experience

6.1.1 Horns Rev I wind farm, Denmark

The first large offshore wind farm in the world, consisting of 80 Vestas V80-2 MW WTs each of 5027 m^2 swept area, was completed in 2002 in 6–14 m of water in the North Sea at Horns Rev, 14–20 km off the West Jutland coast of Denmark.

The project was managed by the West Danish utility Elsam, now DONG Energy, and the wind farm is connected via an offshore substation using 30 kV AC collector cables and to shore from the substation via a 150 kV AC export cable. Maintenance of the wind farm was conceived on the basis of using helicopters access to individual WT nacelles via specially designed access platforms built onto the nacelles to accommodate drops and lifts from the Eurocopter EC135.

There were many difficulties in the early operation of the wind farm arising from the process of installing and commissioning the WTs and from the use of the V80 WT, which had previously been used largely onshore. Table 1.2 demonstrates the scale of the challenge undertaken by the industry at Horns Rev.

The problems arose from some aspects of the WT and wind farm design, as follows:

- WT dry-type transformers, installed in the WT nacelles, experienced winding failures due to seismic vibration fretting
- Vibration and other damage to the DFIG generators
- Gear and bearing damage to the WT gearboxes
- Problems with the WT pitch control systems
- Subsequent problems with the collector and export cable arrays

These difficulties lead to a large number of commissioning and post-commissioning visits to individual WTs and the replacement of some WT gearboxes and generators. The situation worsened and the entire fleet of 80 off V80 nacelles were returned to shore for full refurbishment, although this drastic decision was almost certainly facilitated by the proximity of Horns Rev 1 to the Vestas manufacturing plants and the fact that the manufacturer, developer and operator were of the same nationality.

However, many lessons were learnt from this early large offshore wind farm experience.

6.1.2   Round 1 wind farms, the United Kingdom

The following four UK Round 1 wind farms are considered:

- North Hoyle, operational July 2004, 30 Vestas V80-2 MW WTs of 5027 m^2 swept area, in 7–11 m water depth, 9.2 km offshore in the Irish Sea, operated by RWE Npower Renewables
- Scroby Sands, operational January 2005, 30 Vestas V80-2 MW WTs of 5027 m^2 swept area, in 5–10 m water depth, 3.6 km offshore in the North Sea, operated by E.ON Climate Renewables
- Kentish Flats, operational January 2006, 30 Vestas V90-3 MW WTs of 6362 m^2 swept area, in 5 m water depth, 9.8 km offshore in the English Channel, operated by Vattenfall
- Barrow, operational July 2006, 30 Vestas V90-3 MW WTs of 6362 m^2 swept area, in 15–20 m water depth, 12.8 km offshore in the Irish Sea, operated by Centrica/DONG Energy

It is clear from Table 1.2 and the information above that these four, smaller offshore wind farms were less challenging in location than the larger Horns Rev 1. However, the distance offshore and water depth at Barrow were similar to Horns Rev 1.

It is also clear that considerable experience had been gained between 2002 and 2006 deploying, commissioning and operating offshore wind farms.

All the difficulties recorded at the UK Round 1 wind farms, with the possible exception of pitch system problems on V90 WTs, replicate those of Horns Rev 1 experiences, although they seem to be of lesser magnitude and no complete nacelles had to be replaced, and there was a learning curve operating the Vestas V90 WTs.

The operational problems at the four wind farms were set out in Reference 1 and are summarised from the published operational reports as follows:

Scroby Sands (V80s)
In 2005, there was substantial unplanned work attributed to minor commissioning issues, corrected by remote turbine resets, local turbine resets or minor maintenance work, mostly resolved within a day. A smaller number of unplanned works involved larger-scale plant problems with more serious implications, the primary cause being gearbox bearings.

In 2005, 27 generator side intermediate speed shaft bearings and 12 high-speed shaft bearings were replaced. A number of reasons for the gearbox bearing damage were identified related to the bearing designs.

In 2005, four generators were replaced with generators of alternative design.

In 2006, unplanned work involved three outboard intermediate speed shaft gearbox bearings, nine high-speed shaft gearbox bearings and eight generator failures. Generating capacity was also significantly reduced for 2 months when one of the three transition joints in the cable to the beach failed.

In 2007, problems experienced with the generators were resolved by replacing all original generators with a generator of proven design. The gearbox bearing issue was managed in the short term by proactive replacement of the outboard intermediate speed bearings; in addition 12 high-speed shaft bearings were identified as worn during routine internal inspections and proactively replaced before failure. Three gearboxes were also identified as requiring replacement. Capacity was also affected by a transition joint failure in another cable to the beach; commissioning tests also identified a fault in the sub-sea portion of the cable, for which replacement was planned for spring 2008.

North Hoyle (V80s)

In 2004–2005, unplanned work involved a high-voltage (HV) cable fault, generator faults associated with cable connections and SCADA electrical faults.

In 2006, the following issues arose:

- Two generator bearing faults
- Six gearbox faults
- An unplanned grid outage
- Preparation and return of turbines to service further extended downtime
- Downtime owing to routine maintenance and difficulties in the means of access to the turbines

In 2007, the following issues arose:

- Four gearbox bearing faults and chipped teeth resulting in gearbox replacements delayed by the lack of a suitable maintenance vessel
- Two generator rotor cable faults
- Two circuit breaker failures
- One cracked hub strut
- One turbine outage for yaw motor failures
- An unplanned grid outage
- Again downtime owing to difficulties in the means of access

Kentish Flats (V90s)

In 2006, there was substantial initial unplanned work attributed to minor commissioning issues corrected by remote turbine resets, local turbine resets or minor maintenance work.

Other unplanned work involved larger-scale plant problems and included

- main gearbox;
- generator bearings;
- generator rotor cable connections from the slip-ring unit;
- pitch system.

The generator bearing and rotor cable problems were prolonged as the generator sub-supplier undertook the repairs to avoid jeopardising the warranty.

The first main gearbox damage was detected in late 2006 and an intensive endoscope campaign revealed that 12 gearboxes required exchange. In 2007, all 30 gearboxes were exchanged owing to incipient bearing failures in the planetary gear. The exchange programme was scattered over the year, and due to waiting time and the lack of a crane ship, the outages were longer than the repair time. About half of the generators were refurbished owing to

- damage on internal generator rotor cable connections;
- shaft tolerances;
- grounding of bearings to avoid current passage.

  Other unplanned tasks included

- pitch system repair;
- blade repair on one turbine due to crane impact during gearbox exchange.

Barrow (V90s)
In 2006–2007, unplanned work on the turbines was substantial although some issues were minor, solved by a local reset or minor work to the turbine. Other larger issues were

- generator bearings failed and replaced with a new type;
- generator rotor cables replaced with a new type;
- pitch systems modified.

Owing to gearbox problems seen on other turbines of the same type, an inspection process commenced in 2007 showing a few gearboxes beginning to show similar problems. It was decided proactively to replace gearboxes before failure and this started in July 2007 completing in October 2007.

6.1.3   *Egmond aan Zee, Netherlands*

The Egmond aan Zee wind farm in the Netherlands consists of 36 Vestas V90-3 MW WTs of 6362 m^2 swept area, in 17–23 m water depth, 10–18 km offshore in the North Sea, went operational in April 2007. The wind farm is operated by NoordzeeWind, a joint venture between utility company Nuon, now part of Vattenfall, and Shell.

The location of Egmond aan Zee can be considered to be as challenging as Horns Rev 1; however, Egmond aan Zee has an operational advantage in being so close to a maritime centre at the mouth of the River Ij at Ijmuiden.

Reliability analysis for the Egmond aan Zee wind farm used information taken from operational reports in its first 3 years of operation recording 'stops' and not 'failures' [1, 2], and the results are shown in Figure 6.1.

It can be seen from Figure 6.1(a) that there were a significant number of stops associated with the turbine control system; however, the average downtime per stop data (Figure 6.1(b)) shows that the control system stops must be easily reset as the downtime was short.

Figure 6.1(b) shows high gearbox and generator downtimes, although these components have a relatively low stop rate. This combination leads to very high average downtime per turbine per year. This is confirmed in Figure 6.1(c) where the gearbox is shown to have a significant effect on the turbine availability. The V90-3 MW turbines have been subject to an extensive gearbox replacement programme due to type faults and some generator replacements, reflecting the experience at Kentish Flats and Barrow in the United Kingdom, which also operated the V90-3 MW WTs.

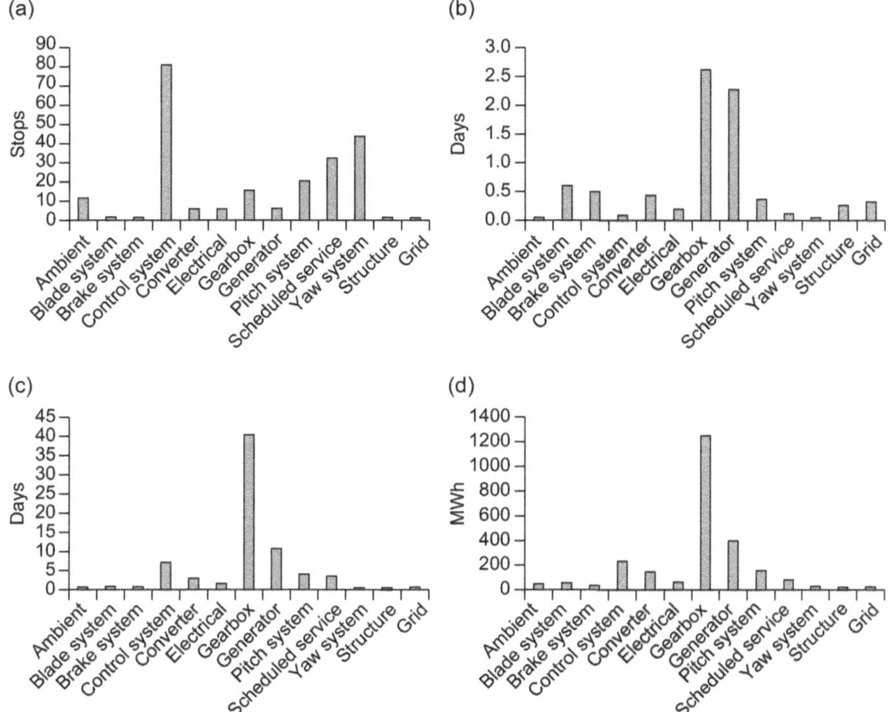

*Figure 6.1   Reliability data for Egmond aan Zee wind farm 2007–2009:
(a) Average number of stops per turbine per year; (b) average downtime
per stop; (c) average downtime per turbine per year; (d) average
energy lost per turbine per year [Source: NoordzeeWind [1, 2]]*

These programmes will have had a significant effect on the downtime figures.

As expected, the average energy lost per turbine per year (Figure 6.1(d)) is closely correlated with the average downtime per turbine per year.

6.2   Experience gained

6.2.1   General

The information on early experience was available from offshore wind farms operating Vestas WTs, information on other makes of WTs, particularly Siemens, is now also becoming available, see Section 6.2.5.

A most interesting conclusion from Horns Rev 1, UK Round 1 and Egmond aan Zee offshore experiences is that the failure modes do not seem markedly different to those described in Chapters 3 and 4 from onshore WTs. There seem to be few, new, unexpected failure modes associated with the offshore environment, except those due to the offshore AC connector cable arrays and the AC export cables.

There were few problems with blades and a large number of problems associated with gearboxes, generators, pitch systems and the turbine control, almost certainly aggravated by the relatively low offshore operational experiences with these two makes of WT.

However, it is clear that the root causes of failure were exacerbated by offshore operations and conditions, for example, due to

- high wind resource;
- consequent drive train transient torques arising from that resource variability;
- WT control system operation;
- seismic vibration of drive trains.

In addition, it is clear that the most pressing issue for all these wind farms was that of access.

6.2.2   Environment

The effect of the offshore environment can be most clearly seen by comparison with the effect of wind speed on the availability of a large onshore wind farm in the United States over a period of 2 years (Figure 6.2).

Then consider in Figure 6.3 the effect of the offshore environment, in particular wind speed to the same scale as Figure 6.2, on the capacity factors of five of the offshore wind farms considered above. Figure 6.3 shows the much increased range of wind speed available to the offshore wind farms but a drop in capacity factor as wind speed rises, although this is less marked for one wind farm.

The importance of achieving high availability at high wind speeds offshore is exemplified by Figure 6.4, taken from [1, 3], which shows availabilities from a large database of onshore WTs and confirms from the energy curve that 40% is available at wind speeds >11 m/s. In References 1 and 4, wind farm availability has also been considered.

It is not clear whether the drop in capacity factor is due to increased outages from higher wind speeds or the fact that already defective WTs cannot be repaired

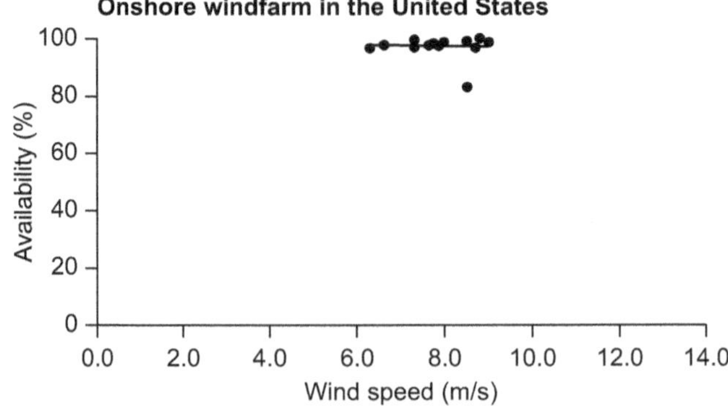

Figure 6.2   Effect of wind speed on WT capacity factor from a large onshore over a period of 1–2 years

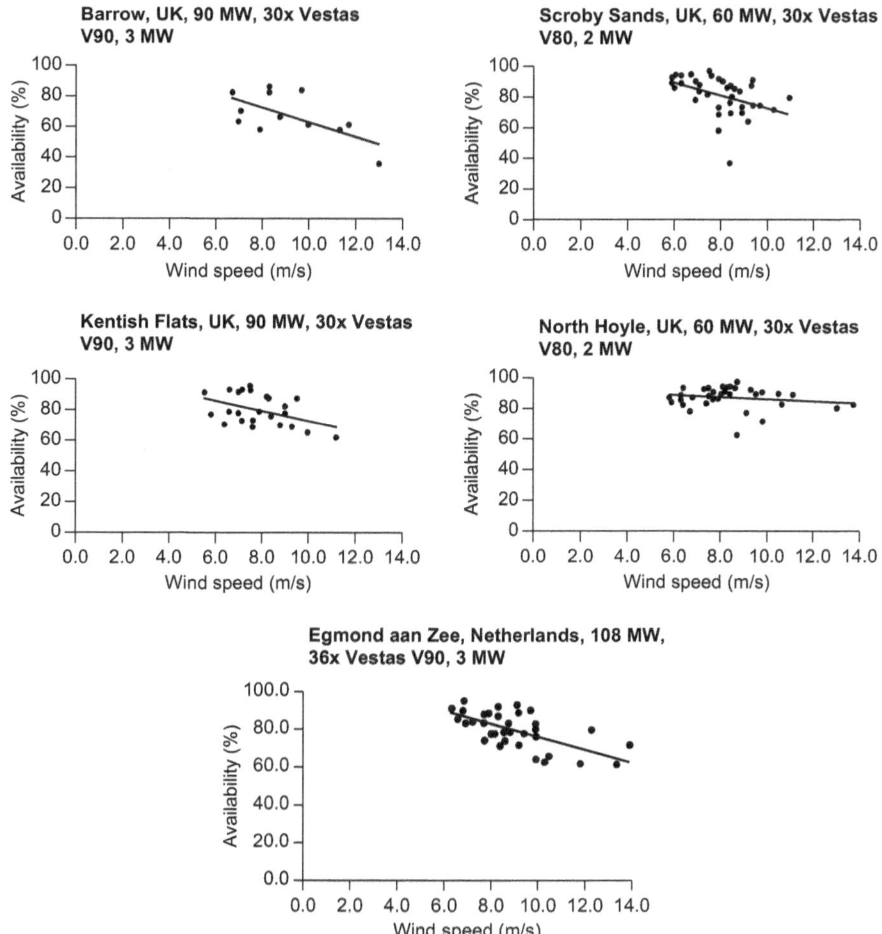

Figure 6.3   Effect of wind speed on WT capacity factor from five offshore wind farms over periods of 1–2 years

at higher wind speeds due to limited access. It is likely to be a combination of both.

These data from five of the offshore wind farms described above can then be compared in Figure 6.5 with the predicted capacity factors for the V80 and V90 WTs. It is clear that both types of WTs are performing reasonably well compared to their theoretical capability, with some fall off at higher wind speeds, which deserves further investigation.

The overall environmental effects are summarised from 2004 to 2009 in Figure 6.6, showing the average wind speed, capacity factor and availability for North Hoyle, Scroby Sands and Egmond aan Zee offshore wind farms, situated

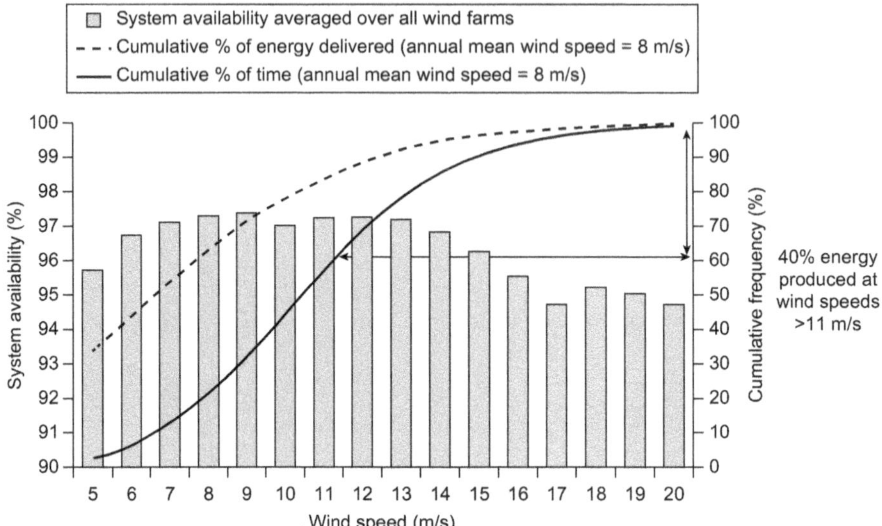

Figure 6.4   Availability of onshore WTs showing the fall in availability at higher wind speeds [Source: GL Garrad Hassan [1, 3]]

respectively in the Irish Sea, North Sea west coast and North Sea east coast. The high capacity factore and low availability coincdes with the operational winter seasons, October to March.

Figure 6.6 shows that the average monthly wind speeds at the three sites were similar, but capacity factors peaked in the highlighted winter seasons, due to higher wind speeds, while availability dipped during those periods. However, close observation of Figure 6.6 shows that at North Hoyle, where Feng *et al.* [1] and perhaps Figures 6.3 and 6.5 suggested an emphasis on O&M, those winter dips in availability were not so severe. This implies that if maintenance and repair are appropriately managed, it is possible to profit from the strong winter capacity factors without a drastic fall in availability. However, these issues depend upon good access and planned maintenance.

There are few records of environmental failure modes, due to corrosion or marine fouling, from the wind farms considered above, although these were well known from earlier, older offshore wind farms. Perhaps these issues will arise as the wind farm life progresses but do not appear to be root causes in the first 3 or 4 years of wind farm life.

6.2.3   Access

The access problems recorded, exemplified in Figure 6.4 in the winter periods by the lower availabilities, were all especially severe when dealing with major repairs, such as the changing of generators and gearboxes, leading to large delays and consequent loss of generated energy.

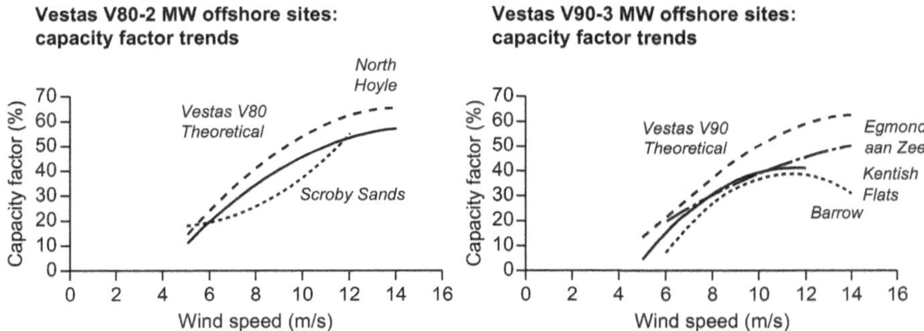

Figure 6.5   Predicted capacity factors for Vestas V80 and V90 WTs compared with measured values achieved at five offshore wind farms

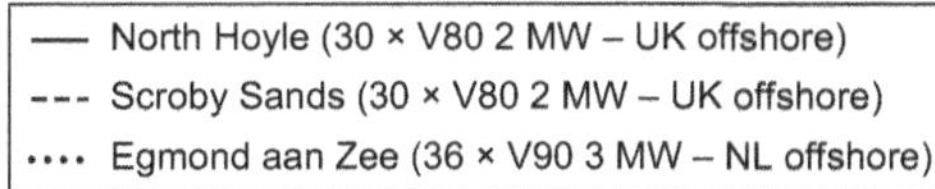

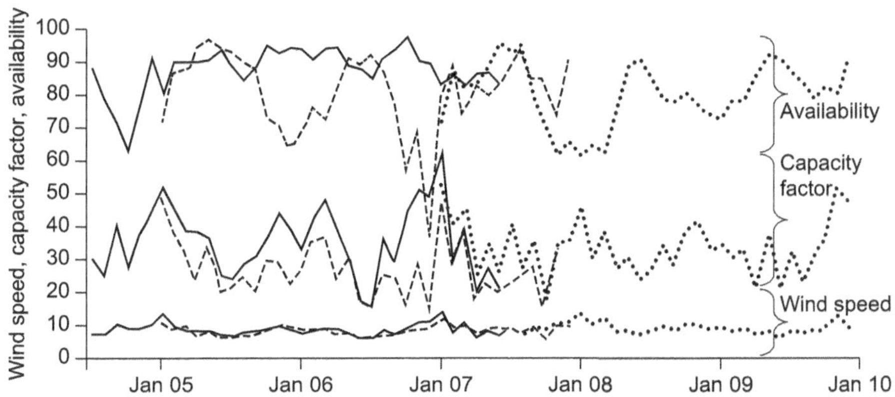

Figure 6.6   Summary of three offshore wind farms with Vestas WTs in early operation

These access issues were not necessarily mitigated by the use of helicopter access at Horns Rev 1, which proved costly and involved personnel difficulties. However, it is believed that these difficulties can be resolved and many recent offshore WTs are now fitted with helicopter drop and lift access platforms (see Figure 5.10) like those used at Horns Rev 1.

6.2.4   Offshore LV, MV and HV networks

6.2.4.1   Substation

The offshore substation at Horns Rev 1 proved to be a success.

6.2.4.2   Collector cables

One UK Round 1 site experienced problems with a collector cable array and other sites experienced problems with some collector cables.

Some of these difficulties arose because buried cables became exposed and then subject to damage from fishing or anchor activity. But some difficulties have arisen because of damage to collector cables due to subsequent construction activity on the wind farm site owing to the activity of jack-up vessels.

6.2.4.3   Export cable connection

Only Horns Rev 1 had an HV export cable and this did experience some problems.

One UK Round 1 site experienced problems with a transition joint in the cable coming to shore.

6.2.5   *Other Round 1 wind farms, the United Kingdom*

Whilst Figure 6.6 shows the performance of three wind farms with Vestas WTs, there is relatively little data in the public domain about other WT makes. However, UK Round 1 and 2 are now operating with Siemens SWT 3.6 WTs with an induction generator drive and fully rated converter, Type D in Figure 4.2. Capacity factor information is available for at least four of these wind farms, summarised as follows, with results shown in Figure 6.7:

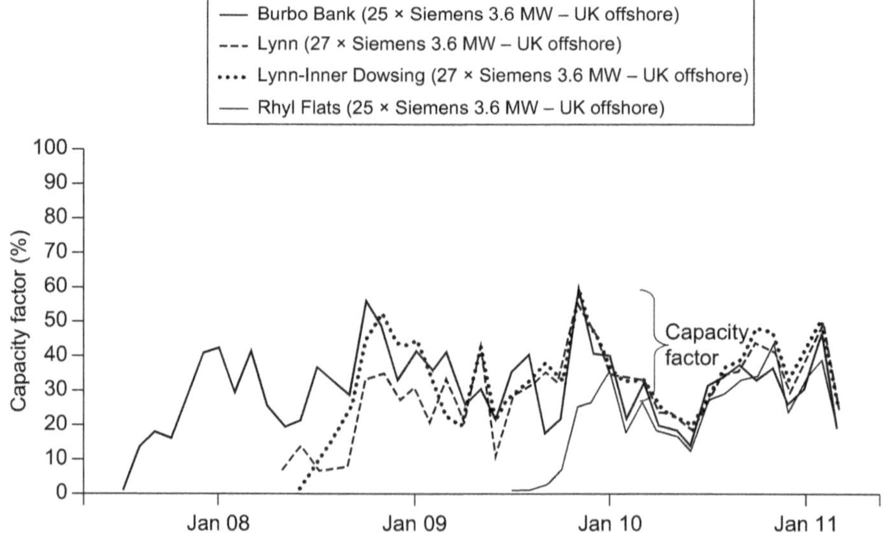

Figure 6.7   Summary of four UK offshore wind farms with Siemens 3.6 MW WTs in early operation

- Burbo Bank, operational July 2007, 25 Siemens SWT 3.6 107-3.6 MW WTs of 9000 m^2 swept area, in 0–8 m water depth, 7 km offshore in the Irish Sea, operated by DONG Energy
- Lynn, operational April 2008, 27 Siemens SWT 3.6 107-3.6 MW WTs of 9000 m^2 swept area, in 5–10 m water depth, 5.2 km offshore in the North Sea, operated by Centrica
- Lynn-Inner Dowsing, operational June 2008, 27 Siemens SWT 3.6 107-3.6 MW WTs of 9000 m^2 swept area, in 5 m water depth, 5.2 km offshore in the North Sea, operated by Centrica
- Rhyl Flats, operational July 2009, 25 Siemens SWT 3.6 107-3.6 MW WTs of 9000 m^2 swept area, in 4–15 m water depth, 8 km offshore in the Irish Sea, operated by RWE Npower Renewables

Figure 6.7 shows capacity factors rising to high values in the winter seasons, similar to Figure 6.6. A most interesting feature is how performance of the wind farms improves progressively in the first 3 years of operation, exemplified by all four offshore wind farms in this case. This is the result of effective early commissioning.

6.2.6   Commissioning

High-quality commissioning proved an important early experience lesson from off-shore wind farms for Horns Rev I, UK Round 1, and at Egmond aan Zee and is exemplified by the improving capacity factors during early operation seen in Figure 6.7.

A notable feature mentioned in a number of the early operating reports was that many early faults were corrected by remote or local turbine resets and that there were a number of SCADA electrical faults, remote and local turbine resets also had to be adjusted to ensure reliable operation.

6.2.7   Planning offshore operations

To plan offshore O&M and avoid the issues that arose in our early European experience, an approach needs to be adopted for operations and maintenance similar to that proposed in Figure 5.3 for design and manufacture. This is shown in Figure 6.8, and the basis is to link that work to the planning of maintenance through the reliability-centred maintenance plan. Again, the key to success for Figure 6.8 is that O&M must soundly base on data generated from the wind farm in operation, just as Figure 5.4 was based upon data generated during design and manufacture.

6.3   Summary

This chapter has shown Europe's early experience of operating offshore wind farms. The experience available in the public domain has been limited to wind farms operating Vestas V80 and V90 WTs but has identified some important general lessons.

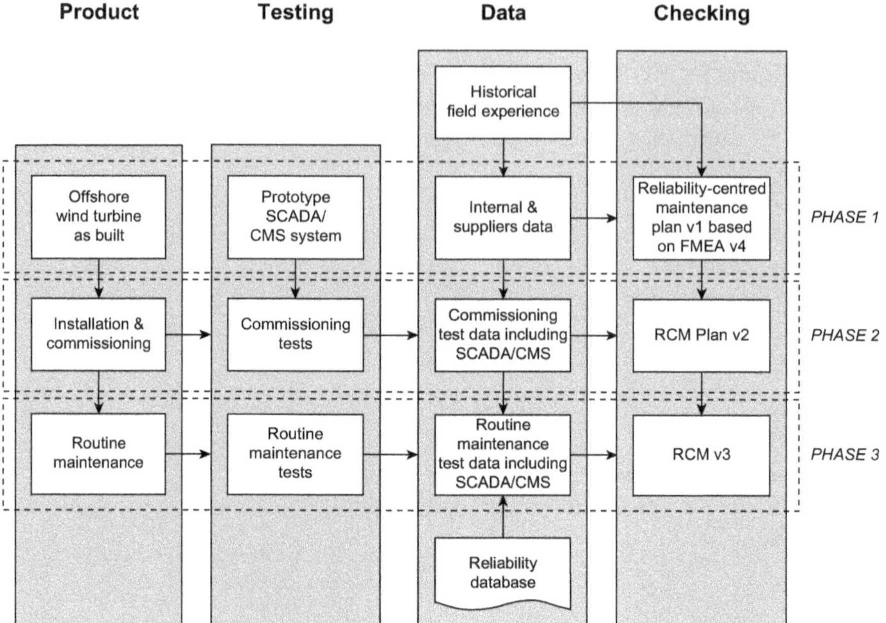

Figure 6.8   Proposal for using FMEA and RCM as OWT review tool for O&M

The major lessons learnt were as follows:

- Onshore WTs do experience problems in the offshore environment; however, many of the failure modes offshore were similar to those that arose in onshore experience.
- Thorough pre-testing of the sub-assemblies and of the WT, designed for offshore operation, are necessary preliminaries to de-risking offshore operation.
- Thorough and efficient commissioning of the WTs in the offshore wind farm lowers the risk of subsequent problems.
- Thorough preparations of offshore access facilities, both at the shore base and wind farm, are essential de-risking preliminaries to offshore O&M.

The next chapter will show how SCADA and CMS monitoring can assist in producing the data to solve the problems of this early experience and turn WT reliability into wind farm availability and a lower cost of energy.

6.4   References

[1]   Feng Y., Tavner P. J., Long H. 'Early experiences with UK round 1 offshore wind farms'. *Proceedings of Institution of Civil Engineers, Energy.* 2010;**163**(EN4):167–81

[2]   NoordzeeWind Various Authors:
   a. *Operations Report 2007*, Document No. OWEZ_R_000_20081023, October 2008. Available from http://www.noordzeewind.nl/files/Common/Data/OWEZ_R_000_20081023%20Operations%202007.pdf?t= 1225374339 [Accessed January 2012]
   b. *Operations Report 2008*, Document No. OWEZ_R_000_ 200900807, August 2009. Available from http://www.noordzeewind.nl/files/Common/Data/OWEZ_R_000_20090807%20Operations%202008.pdf [Accessed January 2012]
   c. *Operations Report 2009*, Document No. OWEZ_R_000_20101112, November 2010. Available from http://www.noordzeewind.nl/files/Common/Data/OWEZ_R_000_20101112_Operations_2009.pdf [Accessed January 2012]

[3]   Harman K., Walker R., Wilkinson M. 'Availability trends observed at operational wind farms'. *Proceedings of European Wind Energy Conference, EWEC2008*. Brussels: European Wind Energy Association; 2008

[4]   Castro Sayas F., Allan R. N. 'Generation availability assessment of wind farms'. *IEE Proceedings Generation, Transmission and Distribution*, Part C. 1996;**1043**(5):507–18

Chapter 7

Monitoring wind turbines

7.1 General

The monitoring of modern WTs may include a variety of systems as follows:

- Supervisory Control and Data Acquisition (SCADA) system, to provide low-resolution monitoring to supervise the operation of the WT and provide a channel for data and alarms from the WT.
- Condition Monitoring System (CMS), to provide high-resolution monitoring of high-risk sub-assemblies of the WT for the diagnosis and prognosis of faults, included in this area are Blade Monitoring Systems (BMS), aimed at the early detection of blade defects.
- Structural Health Monitoring (SHM), to provide low-resolution signals for the monitoring of key items of the WT structure.

These systems each have different data rates and summarised in Figure 7.1, as the wind industry develops they are slowly being integrated together.

7.2 Supervisory Control and Data Acquisition

7.2.1 Why SCADA?

Supervisory Control and Data Acquisition (SCADA) systems originated in the oil, gas and process industries where large physically distributed processes could only be controlled by accurate measurements of the status of valves, pumps and storage vessels and of the consequent temperatures, pressures and flows.

These data acquisition systems were originally developed independent of the controls. However, where their measurements were needed to control the plant, they evolved into Industrial Control Systems (ICS). More recently, where plant control was distributed throughout the plant and embedded into the data acquisition system, some SCADA systems evolved further into Distributed Control Systems (DCS).

The power generation industry has been using SCADA for 35 years and DCS has been used to control modern power station units in the United Kingdom since about 1985. Therefore, it was natural for the wind industry to apply these techniques to the WT, an unmanned remote robotic power generation unit.

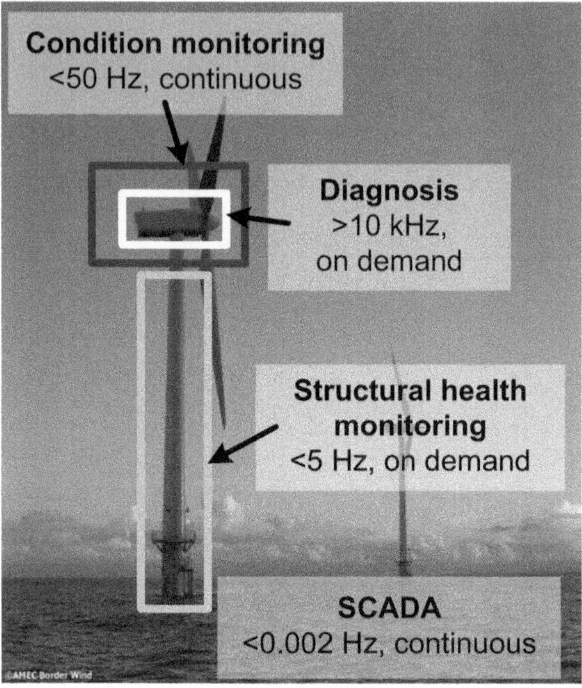

Figure 7.1   SCADA and CMS of a WT

However, the emphasis in the wind industry has been on monitoring rather than control, which in a WT is exercised primarily by the WT controller mounted in the nacelle, although that can be overrode by external signals from the operator via SCADA.

In fact, the majority of SCADA signals and alarms derive from within the WT controller, which is generally an industrial programmable logic controller (PLC) that ensures that the WT remains within its safe operating envelope supervising cut-in, synchronisation, adherence to the power curve, cut-out and emergency stop action in the event of untoward operation.

The international standard that prescribes the layout for WT communication systems, including SCADA, is given in References 1 and 2 (see Figure 7.2).

However, the evolution in WT size, number of units and designs has encouraged the wind industry to apply SCADA more widely than elsewhere in the power generation. This may have been because of the growth and cheapness of measurement and information and communications technology, but is also because of the prototype nature of early large WT development, exemplified by the latter part of the table in Appendix 1.

A survey of the SCADA systems available to the wind industry is given in Chapter 13, Appendix 4.

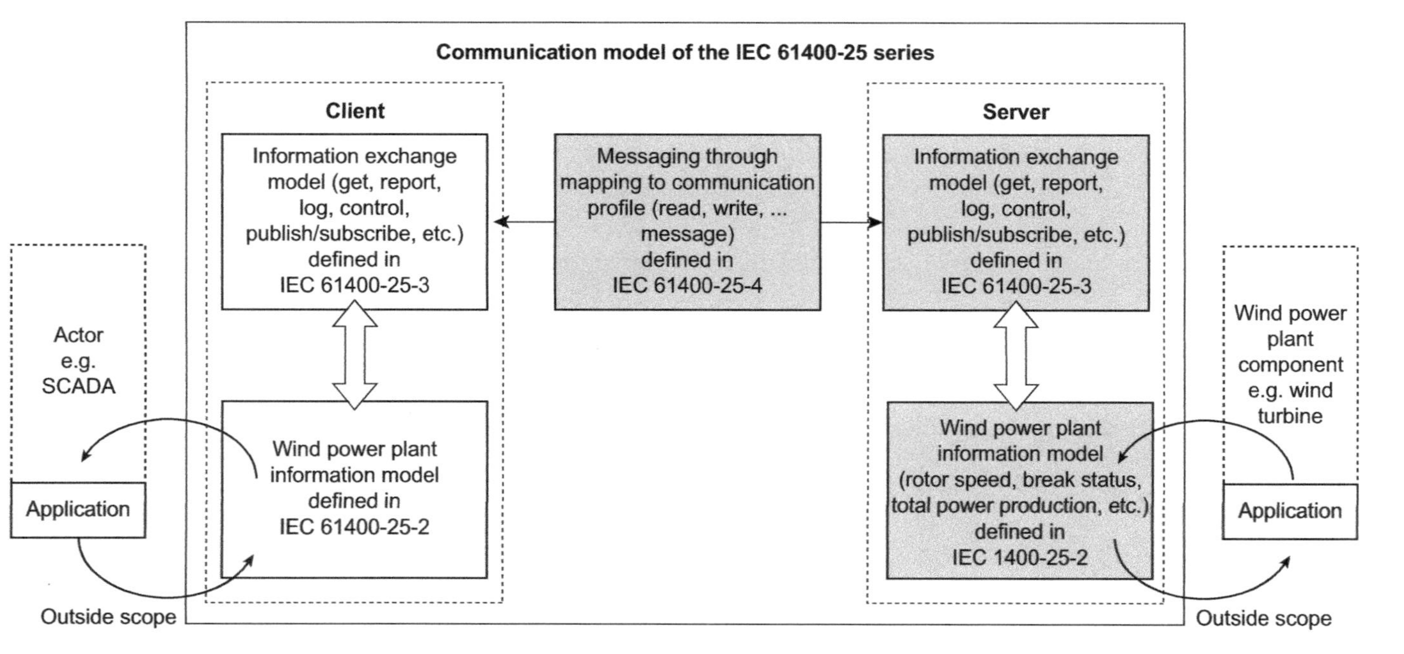

Figure 7.2   Conceptual communication model for a WT [Source: [1]]

7.2.2   Signals and alarms

The SCADA system handles both input/output (I/O) signals and alarms and usually samples signals at 10 minute intervals, although for fast changing or commercially valuable signals, such as wind speed or power output, systems can record and transmit maximum, mean, minimum and standard deviations of the signal.

The majority of data are output, flowing from the WT to the control room, but some signals and commands are input, fed from the control room to the WT.

To give an example of the growth of modern wind industry SCADA, an operational 500 MW fossil-fired generation unit may have 1–2000 SCADA I/O channels, whereas a modern 5 MW offshore WT may have 4–500 I/O channels, including signals and alarms, emphasising the unmanned, remote, robotic and developmental nature of modern, large WT units.

7.2.3   Value and cost of SCADA

The value of SCADA is that it gives the WT OEM, or operator, online data about the functioning and alarms of WTs remote from their operational base. This allows the generation of graphical information to allow operations to be optimised and maintenance to be planned, for example see Figure 7.3.

However, the volume of data generated by SCADA requires careful organisation, for example an offshore wind farm with 100 WTs each generates 40,000 data

Figure 7.3   Analysing SCADA data to detect wind turbine problems [Source: GL Garrad Hassan]

items every 10 minutes, that is, 96 MB of data per day, requiring considerable analysis for online interpretation.

In general, WT OEMs have developed these techniques in order to manage WTs during the warranty period and, using SCADA, are able to compare the performance of different wind farms and the performance of individual WTs against the whole populations of that type.

A great benefit of SCADA is that it provides an overview of the whole WT, looking at production measures, such as wind speed and energy output, monitoring signals, such as lubrication oil and bearing temperatures and control system alarms from the pitch and power electronics systems, for example. Therefore, it can allow the operator to compare signals widely across the WT system giving confidence in indicated results. The weakness of SCADA is that its low data rate does not allow the depth of analysis that is usually associated with accurate diagnosis. However, as the next section will show, this weakness in depth is more than compensated by the breadth of SCADA scope, which can produce easy-to-interpret graphical images, such as the power curves shown in Figure 7.3.

On the other hand, WT operators generally do not have these facilities, except by access permission from the WT OEM, and face difficult decisions at the end of the warranty period, whether to extend an OEM maintenance contract or attempt to manage the WTs themselves.

SCADA is generally a low-cost monitoring system, integrating cheap, high-volume measurement, information and communications technology into the WT controller by the OEM during original manufacture. A typical cost of SCADA provision depends upon the size of a wind farm but can be typically £5000–10,000/WT.

7.3   Condition Monitoring Systems

7.3.1   Why CMS?

WT CM first appeared in the industry in the 1990s, following pressure from insurance companies, as a reaction to a large number of claims due to high-profile WT gearbox failures, and the technology was largely adapted from other rotating machine vibration CM experience. WT CMS initially came from reputable condition monitoring OEMs, such as Bruel and Kjaer, Bently Nevada and National Instruments, and the systems were largely based on experience in traditional rotating machine vibration condition monitoring experience and their selection became part of the WT certification process [3].

However, WT condition monitoring throws up a number of issues, which are not common in traditional rotating machines, based on the stochastic nature of the wind resource, that is the modern large WTs operate at continuous and rapidly varying power, torque and speed and are usually remote from technical support.

CMS has proven successful in onshore WTs when used by experienced operators and is now installed on new WTs $\geq$ 1.5 MW almost as standard and has been fitted to almost all offshore WTs.

WT OEMs make considerable use of CMS technology during the WT warranty period, but despite the continued installation of WT CMSs, little attention is paid by operators to the alarms and data generated by the systems. This stems primarily from the fact that operators may not have the specialist knowledge required to interpret complex vibration CM data. As a result, many operators, particularly those with less experience, subcontract WT CM to specialist companies or maintain a monitoring contract with the WT OEM. This can be a costly exercise and CMS may, as a result, be neglected and reactive maintenance strategies be adopted.

A survey of the CMS systems available to the wind industry is given in Chapter 14, Appendix 5.

7.3.2   *Different CMS techniques*

7.3.2.1   **Vibration**

Vibration techniques were the first to be used in WT CMS, initially for monitoring the generator, the gearbox and the main bearing of the turbine. A variety of techniques have been used including low-frequency accelerometers for the main bearings and higher-frequency accelerometers for the gearbox and generator bearings and in some cases proximeters. Figure 7.4 shows the frequency range appropriate for vibration displacement, velocity and acceleration measurements.

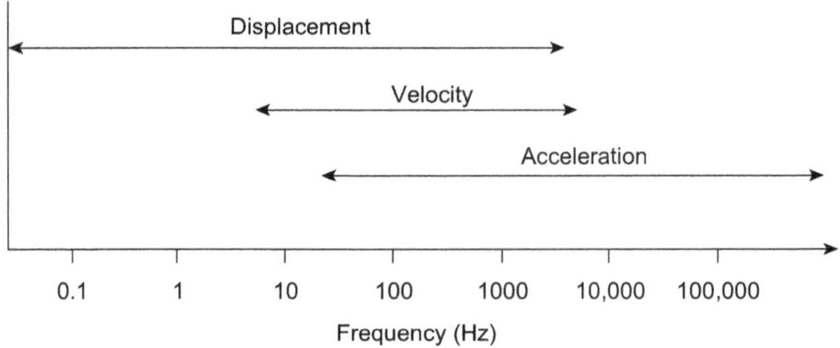

Figure 7.4   Frequency range for displacement, velocity and acceleration vibration measurement

A particular issue for WT vibration analysis is that vibration periods and amplitudes change with time, as a consequence of the continuously and rapidly changing drive train torque, and care is essential during the analysis.

A feature of WT condition monitoring is that the majority of bearings within the drive train are rolling element and that, combined with the use of a high ratio gearbox, means that when faults are present vibration signals contain a high impulsive content.

This and the continuously changing drive train torque have encouraged some to advocate the use of wavelets [3, 4] to analyse WT CMS signals to deal with its time-varying and impulsive nature. However, this is computationally expensive.

The majority of current CMS vibration analyses methods used by the wind industry, described in Chapter 13, capitalises upon the periodic origin of the vibration signals and uses conventional Fourier Transforms, but with the signal is collected within a limited, pre-defined speed and power range.

The most important issues to consider in analysing these vibration measurements are the following:

- Vibration peak and rms amplitude trends
- Vibration signal time domain
- Vibration signal frequency content

A rising rms vibration trend indicates a worsening fault but a low rms vibration, and with high peak value indicates impulsive energy in the signal and the need to observe the time domain to determine waveform content and identify the impulsive component. Finally, if the time domain confirms an unusual impulsive component structure, vibration frequency content analysis is necessary. This can identify the harmonic origins of the impulsive energy content, for example, gear- or ball-passing frequencies, enabling the vibration source to be located. Vibration signals are rich in harmonic information, which must be accurately understood if diagnosis is to be performed with confidence. Some WT CMS systems allow the mechanical parts of the drive train to be represented within the CMS to provide Fast Fourier Transform (FFT) spectral cursors to aid interpretation.

7.3.2.2   Oil debris and analysis

Because of the seriousness of gearbox failures, in terms of downtime, gearbox oil debris analysis has assumed more importance in the industry. The function of the oil in the gearbox is three-fold:

- To provide cooling for the gearbox
- To provide lubrication for the rolling element bearings
- To provide lubrication for the meshing gears

The lubrication oil itself will have base properties to ensure proper formation of the lubrication film in the gear pairs and bearings and have additives to minimise wear. Maintenance of these good lubrication properties depends upon

- a high-quality charge of oil in the first place;
- removal of debris;
- maintenance of the oil at suitable temperature;
- cleaning of the oil at appropriate intervals or renewal with the same grade and quality.

Gears and bearings are all wearing components and inevitably produce some ferrous and non-ferrous debris from their natural operation. Debris produced during the gearbox running in process should have been removed by running in tests during production such as that shown in Figure 5.9. Most WTs, in common with other large gearboxes, for example in the marine industry, utilise a spray lubrication system. That means that oil is pumped from the gearbox sump, via a cooler and in-line filter, to the top of the gearbox, from whence it is sprayed onto the operating components from a number of nozzles. Oil is not therefore introduced directly to the bearing or gear pair to be lubricated via an oil port, as it would be, for example, in an internal combustion engine.

This means that the WT gearbox oil stream is both universal and mixed, gaining heat and debris from all parts of the gearbox. An advantage of this is that any oil monitoring process is inevitably global for the whole gearbox, making it attractive for condition monitoring.

Crucial to the value of oil debris detection is the length of the warning that it can give of impending failure (Figure 7.5), giving time to arrange for inspection and maintenance.

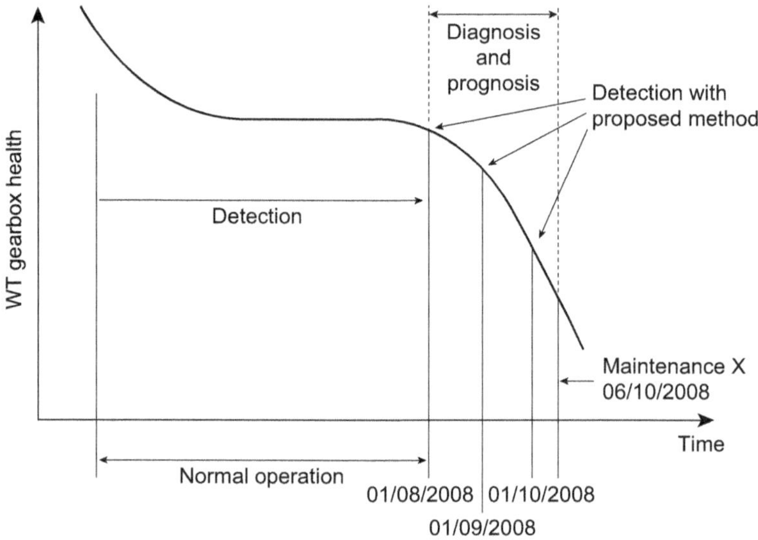

Figure 7.5   Example of the detection process

This latter point is at the heart of effective condition monitoring.

However, oil debris detection cannot then locate a fault, except by distinguishing between the types of debris produced. The arrangement of a three-stage gearbox is showing the location of parts and the oil system diagrammatically in Figure 7.6.

An in-line filter can remove large debris items, >100 μm in diameter, but cannot remove smaller debris without excessive pressure drop. Studies have shown that gearbox oil should be maintained below 2 μm and there are life advantages if it is kept below 1 μm. However, practically few gearboxes can achieve this level of cleanliness. Modern oil debris counters take a proportion of the lubrication oil stream from downstream of the filter in Figure 7.6 and detect and count both ferrous and non-ferrous particles of varying sizes. The counts can be fed as online data to a CMS. Increasing measurement detail increases the cost of the online instrument.

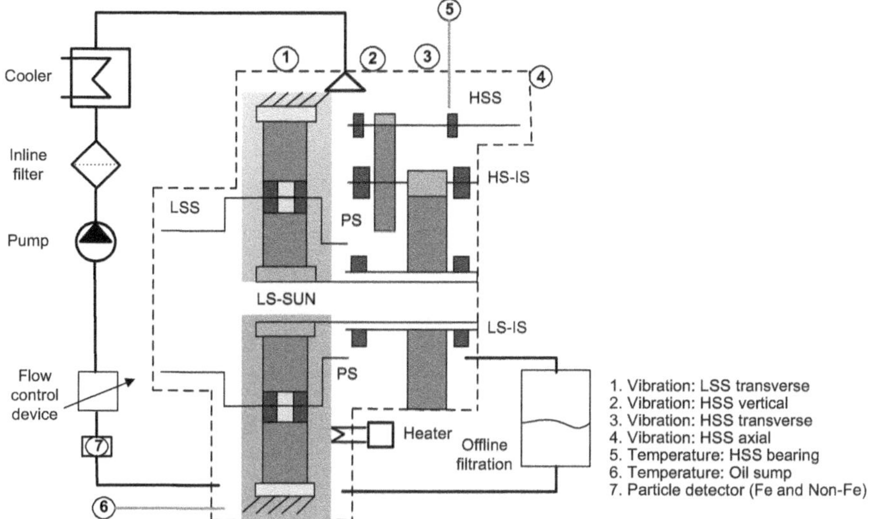

Figure 7.6   Diagrammatic layout of a three-stage gearbox and lubrication system showing measurement points

7.3.2.3   Strain

In order to improve WT performance there has been a trend for variable pitch WTs over the past 5 years to adopt a process of independently pitching the three blades of the turbine. This independent pitch control reduces the torque and lateral force loads on the WT, prolonging life, and is possible through independent blade root bending moment measurements made using circular fibre optics incorporating Fibre Bragg Gratings (FBG) strain gauges, such as that shown in Figure 7.7. The measurements from these strain gauges are primarily intended for blade pitch control. However, these measurements can also now be used to condition monitor the WT performance, and these techniques are growing in the industry, see the survey in Chapter 14.

7.3.2.4   Electrical

Finally, the newest potential source of condition monitoring information from the WT comes from the electrical signals, voltage, current and power used to control the generator speed and excitation. These signals have been used for many years for condition monitoring electrical machines and their coupled drive trains [5]. They can now be used as global monitoring signals for the WT drive train, particularly the power [6]. The difficulty with these electrical signals is that they are very rich in harmonic electrical information, which must be accurately understood if diagnosis is to be performed with confidence [7]. A similar method such as that used to generate CMS FFT spectral cursors for vibration interpretation is needed to aid electrical interpretation.

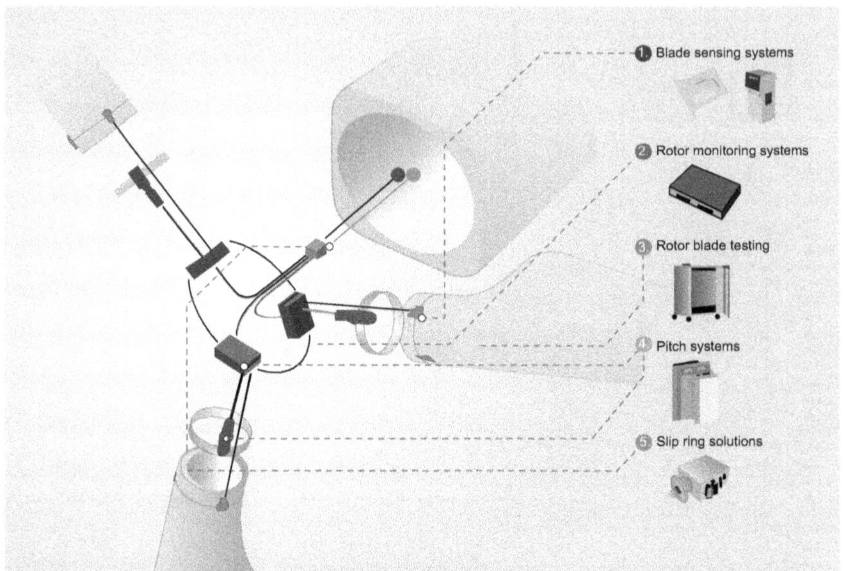

Figure 7.7   Diagrammatic layout of a three-blade fibre FBG pitch control system [Source: Moog Insensys]

7.3.3   Value and cost of CMS

The cost of the hardware and software of a mid-range WT CMS is approximately £7000 and would require approximately £7000 to retrofit to an existing WT, that is, £14,000/WT more expensive than SCADA and with less coverage. These costs would fall if the CMS were installed in volume to a large number of WTS, as is done by the OEM. WT OEMs will wish to fit their own specified CMS, which they have developed over time with their CMS OEM, as this is their main diagnosis tool during the warranty period. Operators may have their own preferences, because of experience with other WT plant, but cannot expect to retrofit their own choice without incurring costs similar to those described above.

Therefore, CMS is not as cheap as SCADA. In addition, CMS data interpretation incurs costs, dependent upon the availability of skilled manpower. There has been considerable debate in the industry about the true value of CMS.

Recent studies by the author have shown that CMS for traditional power generation plant can be justified solely on the saving of costs from unplanned lost production, prevented by the CMS.

For onshore WTs, CMS can be justified, at the cost levels described above, if the costs of replacement equipment, associated labour and lost production are taken into account, particularly if gearbox and generator failures are prevented.

For offshore WTs, CMS can be justified if the costs of site access, replacement equipment, associated labour and lost production are taken into account, again particularly if blade, gearbox, generator or other large sub-assembly failures are prevented.

However, in all these cases, WT CMS can only be justified if the system is capable of detecting a fault and giving sufficient warning (Figure 7.5) to avoid full sub-assembly replacement, the most costly aspect of failure, and if that CMS detection and warning can be acted upon by operators and WT OEMs.

7.4 SCADA and CMS monitoring successes

7.4.1 General

The processes necessary for successful SCADA or CMS monitoring are set out in Figure 7.5, namely:

- Detection, that is the perception that something is faulty in part of the machinery and ideally a location for that fault.
- Diagnosis, that is determination of the nature of the fault, including its more precise location.
- Prognosis, that is determination of what needs to be done to remove the fault.
- Maintenance action, that is to remove the cause of the fault or to replace the faulty item.

A fault will take a certain time to develop before it can interrupt the operation of the WT, and monitoring must consider that time span if it is to be effective. For example, some faults take a short time to grow from inception to failure. A generator earth fault may take 10 seconds to grow from inception to failure. Such a fault may give sufficient time for detection but certainly not for diagnosis, prognosis and maintenance action. On the other hand, Figure 7.5 showed that an oil debris detection process may give some weeks of warning, which if successfully detected by the monitoring system, will allow effective diagnosis, prognosis and potentially successful maintenance action. This period from detection to maintenance outcome has been called the prognostic horizon.

SCADA and CMS monitoring must concentrate on the measurements and detections that can extend this so-called prognostic horizon. The following examples are shown to demonstrate this.

The analysis methods being used to monitor on SCADA and CMS signals include the following:

- Simple trending
- Physics of failure methods
- Narrow band spectral methods
- Fourier transform methods
- Wavelet and non-stationary methods
- Artificial intelligence methods:
  - Artificial neural networks
  - Bayesian methods
- Multi-parameter monitoring

The following successful SCADA and CMS signal or alarm detection of known incipient WT or wind turbine condition monitoring test rig (WTCMTR) sub-assembly faults are taken from referenced examples made by the author's research workers and students using data supplied through our research contracts.

7.4.2   *SCADA success*

The problems of monitoring WTs are clearly demonstrated in Figure 7.8 showing 18 days of SCADA power, wind speed, rotor speed and generator bearing temperature data from a single WT. During this period, a significant storm and high winds were experienced for 2 days, whilst for the remainder the WT operated each day with a diurnal wind speed variation. The reader can see the wide and rapid changes of power and rotor speed, which must be accommodated in any analysis of the signals for detection, diagnosis and prognosis of faults.

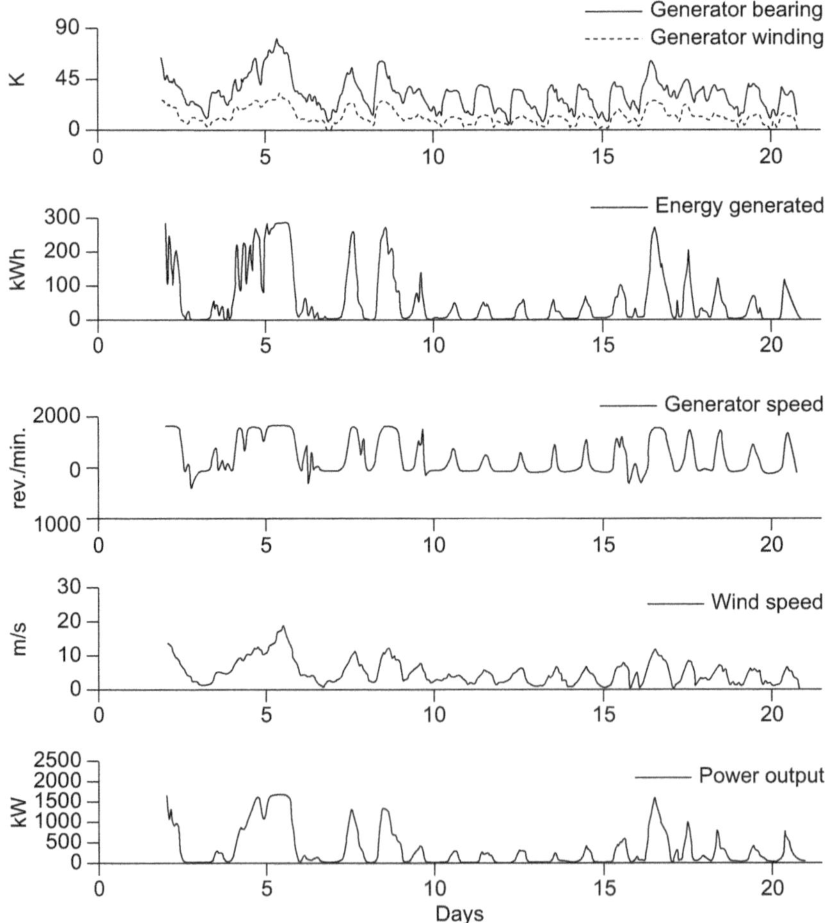

Figure 7.8   18 days SCADA data from a variable speed WT > 1 MW

The most common monitoring analysis applied to WT SCADA data has been to study changes in the WT power curve, as shown in Figure 7.3.

However, more detailed analysis is possible and is exemplified here by studying SCADA signals for two sub-assemblies of concern, the gearbox and the converter.

The first example considers the gearbox, a key WT drive train sub-assembly whose operation differs from conventional generation systems because of the stochastically varying torque experienced in the WT, which is considered to be a major root cause for gear and bearing fatigue. Gearbox root cause analysis requires detailed understanding of the effects of the operating environment and cumulative high- and low-cycle fatigue damage, using information from the gearbox and its neighbouring sub-assemblies, the rotor and generator. This can simply be analysed from SCADA data, using a physics of failure approach, by monitoring the transmission efficiency and rotational speed at different shaft stages and relating them to the gearbox temperature rise to detect and prognose fault development. The heat generated in a gear stage or bearing will be proportional to the work done on that component, which means

$$Q \propto W \propto \Delta T \tag{7.1}$$

Q is the heat generated from the gear stage or bearing, W the work done upon it and ΔT is its temperature rise above nacelle temperature. The work done by a gear stage can be physically expressed as

$$W = \frac{1}{2} I \omega^2 \tag{7.2}$$

Supposing the gear efficiency is η_{Gear} and the bearing has efficiency η_{Brg}, the energy dissipated will be transferred as heat onto the gear or the bearing. Therefore,

$$Q_{Gear} = (1 - \eta_{Gear}) \frac{1}{2} I_{Gear} \omega^2_{Gear} = k_{Gear} \Delta T_{Gear} \tag{7.3}$$

or

$$Q_{Brg} = (1 - \eta_{Brg}) \frac{1}{2} I_{Brg} \omega^2_{Brg} = k_{Brg} \Delta T_{Brg} \tag{7.4}$$

also expressed as the inefficiency $1 - \eta_{Gear}$:

$$1 - \eta_{Gear} = \frac{2 k_{Gear} \Delta T_{Gear}}{I_{Gear} \omega^2_{Gear}} \tag{7.5}$$

or

$$1 - \eta_{Brg} = \frac{2 k_{Brg} \Delta T_{Brg}}{I_{Brg} \omega^2_{Brg}} \tag{7.6}$$

Therefore, $2k/I$ is constant for any gear stage or bearing whose inefficiencies will then be proportional to $\Delta T_{Gear}/\omega^2_{Gear}$ or $\Delta T_{Brg}/\omega^2_{Brg}$, respectively. When a fault, leading to an efficiency reduction, occurs in a gear stage, (7.6) shows that ΔT_{Gear} will increase for the same ω_{Gear}^2.

Assuming that the remainder of the kinetic energy transmitted through gearbox is converted into generator power output such that

$$P_{out} = W - Q_{Gear} \tag{7.7}$$

then

$$P_{out} = \eta_{Gear} \frac{1}{2} I_{Gear} \omega_{Gear}^2 \tag{7.8}$$

By comparing (7.3) and (7.8), we have

$$\frac{1 - \eta_{Gear}}{\eta_{Gear}} = k_{Gear} \frac{\Delta T_{Gear}}{P_{out}} \tag{7.9}$$

or

$$\Delta T_{Gear} = P_{out} \frac{1}{k_{Gear}} \left(\frac{1}{\eta_{Gear}} - 1 \right) \tag{7.10}$$

Equation (7.10) shows that the temperature rise of the gear stage is proportional to the power output P_{out}, given an unchanged gear stage efficiency. At a certain power output, the efficiency η_{Gear} for a healthy gearbox in ideal conditions will be fixed; therefore, ΔT_{Gear} is proportional to power output P_{out}. When a fault occurs in a gear stage, leading to an efficiency reduction, (7.10) shows that ΔT_{Gear} must increase for the same power output P_{out}.

In the following, this approach has been used retrospectively on the SCADA data of a variable speed WT of ~2 MW [11], in which the maintenance record showed a subsequent catastrophic WT gearbox planetary gear failure, undetected by any WT monitoring system. The SCADA analysis has been done on data for three successive identical length periods:

- 9 months before failure
- 6 months before failure
- 3 months before failure

Figure 7.9 shows the average temperature rise ΔT_{Gear} bands plotted against ω_{Gear}^2 and grouped into the separate periods. The data for the 3 month period preceding the failure clearly show the worsening situation predicted by (7.10).

Alternatively, the WT output power can be normalised to the rated power P_N and the gearbox oil temperature rise ΔT_{Gear} assumed proportional to the power out, according to (7.10). Figure 7.10 shows this for the three periods. In this figure, the average gearbox oil temperature rise was binned for 50 kW power output increments in the three periods.

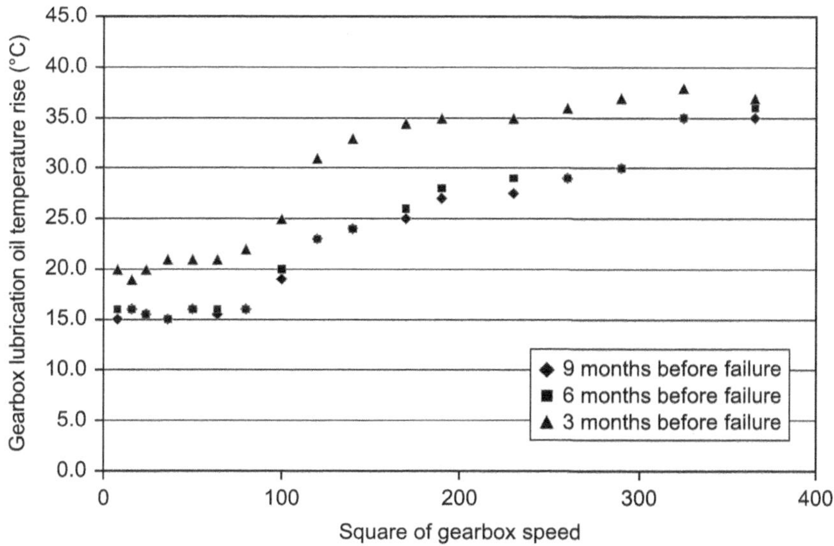

Figure 7.9   *Trends of gearbox oil temperature rise ΔT_{Gear} vs square of rotor velocity ω^2_{Gear}*

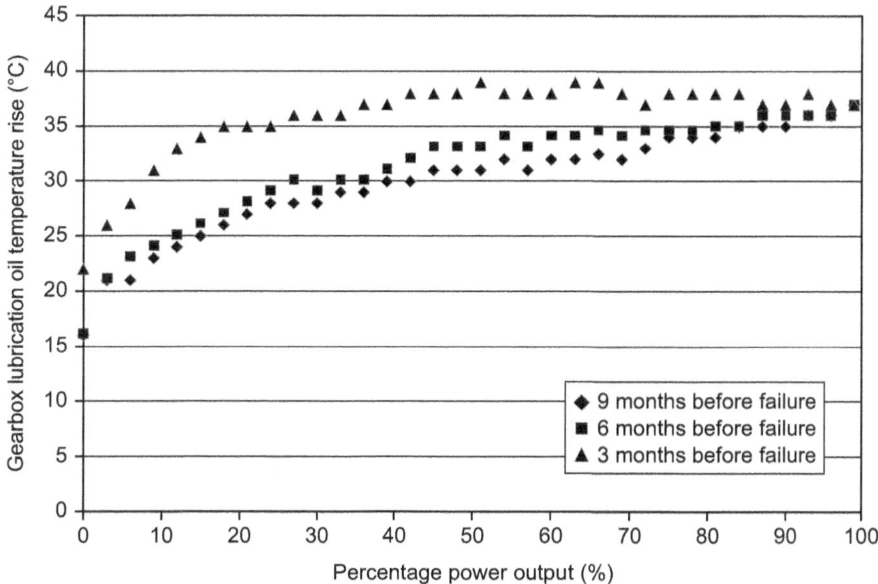

Figure 7.10   *Trends of gearbox oil temperature rise vs relative WT power output*

Figure 7.11 shows the histogram of frequency of gearbox oil temperature rises for the three periods.

Figures 7.10 and 7.11 both clearly show a rising gearbox inefficiency in the 9 months before failure, with a worsening trend in the last 3 months predicting the failure.

The results of Figures 7.9–7.11, conforming to (7.5) and (7.10), demonstrate clearly that slow speed SCADA temperature data can provide long-term detection and prognosis for the internal gearbox problems.

Probably the simplest detection algorithm to adopt, based on Figure 7.11, would be to measure gearbox oil temperature rise and bin results into temperature rise bands, placing an alarm on bands above a 35°C rise.

Others have shown how SCADA monitoring can predict failures, including [8–10].

Another SCADA data monitoring example, intending to predict WT converter sub-assembly failures, investigated alarm showers from WT controller alarm indications [12]. This again adopted a physics of failure approach. To do this the normalised cumulative percentage of selected generator, grid and converter alarms was plotted (Figure 7.12) against calendar time, during an extended period of operation for two specific variable speed WTs of ~2 MW, chosen at random from the same wind farm.

Figure 7.12 shows the following:

- The impact of two grid voltage dip incidents on days 39,200 and 39,500 on the two WT alarm patterns, the same patterns were observed on other WTs in the WF during the same days.

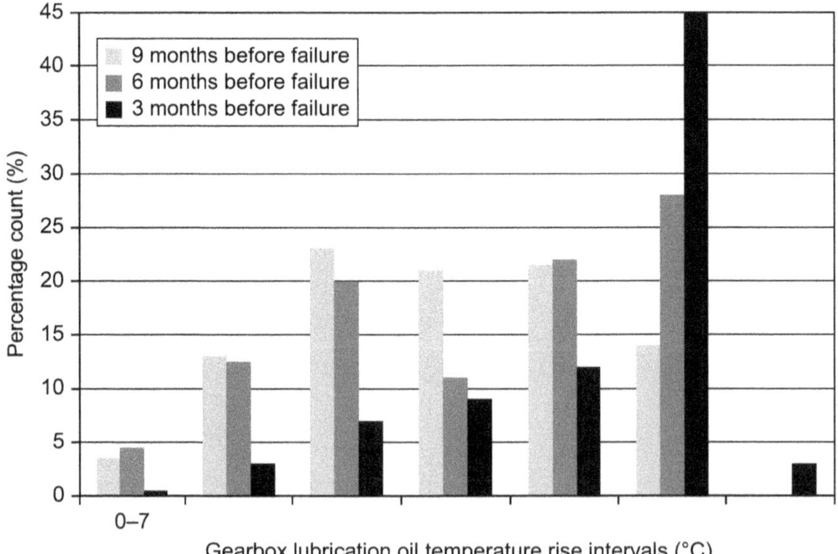

Figure 7.11   Histogram of frequency of gearbox oil temperature rises

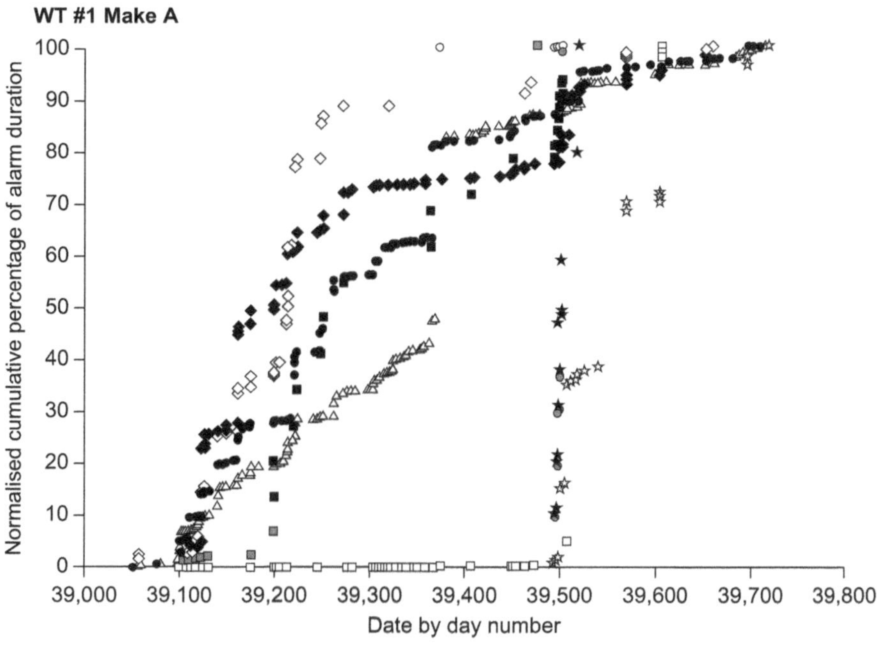

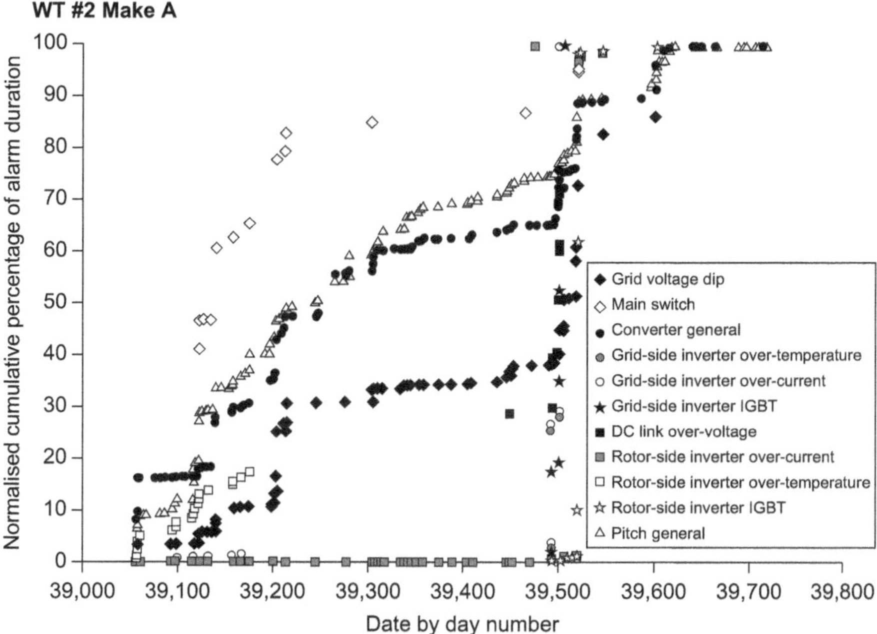

Figure 7.12   Normalised cumulative alarm percentage duration vs calendar time, for two WTs in the same wind farm [Source: [12]]

- Serious grid voltage dips of >75% caused more than 10 converter or inverter alarms during the period investigated.
- Converter general alarms strongly correlated with the grid voltage dip alarms, indicating grid voltage dip as a root cause for converter failures.
- WT EFC pitch alarms also responded to these conditions and their alarms are also shown in Figure 7.12.
- The steps observed in the normalised cumulative alarm percentage indicate alarm triggers with long cumulative duration. The numerous alarms in these steps were accompanied by inverter Insulated Gate Bipolar Transistor (IGBT) failure alarms, suggesting the use of these steps to accumulate inverter sub-assembly stresses, giving advance warning of converter faults.

In total 15–20 alarm triggers were observed for each WT for each of these incidents. Therefore, for a WF with 30–35 WTs, 450–700 alarms could be triggered simultaneously by such incidents. With the probable repetition of some alarms, this suggests a possible alarm rate >1000 per 10 minutes, suggesting a need to optimise WT alarms.

Such simple algorithms could easily be implemented either in the WT controller to give a graph and local alarm or in a remote control centre where the same algorithm could be applied to each WT in a wind farm.

7.4.3   CMS success

The following are examples, using CMS signals, of successful detection of incipient faults in various sub-assemblies from a WTCMTR or operational WTs in the field.

The first examples, using simple narrow band spectral analysis on electrical signals from a variable speed WTCMTR, detected WT DFIG generator rotor electrical unbalance in Figure 7.13 and mechanical unbalance in Figure 7.14. These results are taken from Reference 8.

Figures 7.13 and 7.14 clearly show that WT DFIG electrical or mechanical rotor faults can be detected by simple narrow band spectral analysis of generator CMS electrical signals.

Turning now to the use of CMS on operational WTs in the field, Figure 7.15 shows rising gearbox high-speed stage (HSS) vibration accompanied by a rising oil debris count, indicative of a deteriorating intermediate stage (IMS) bearing. The removed IMS bearing inner race is shown in Figure 7.15 to indicate the damage causing the indications above. The key points to note from Figure 7.15 are

- the combination of indications arising from two disparate sensors;
- the substantial warning obtainable from the measurements, in this case more than 120 days;
- the opportunity to plot data against a variety of variables, see Chapter 2, Section 2.3, may improve detection visibility;
- in this case, timely CMS detection of an incipient IMS bearing fault prevented bearing failure and potential failure of the whole gearbox.

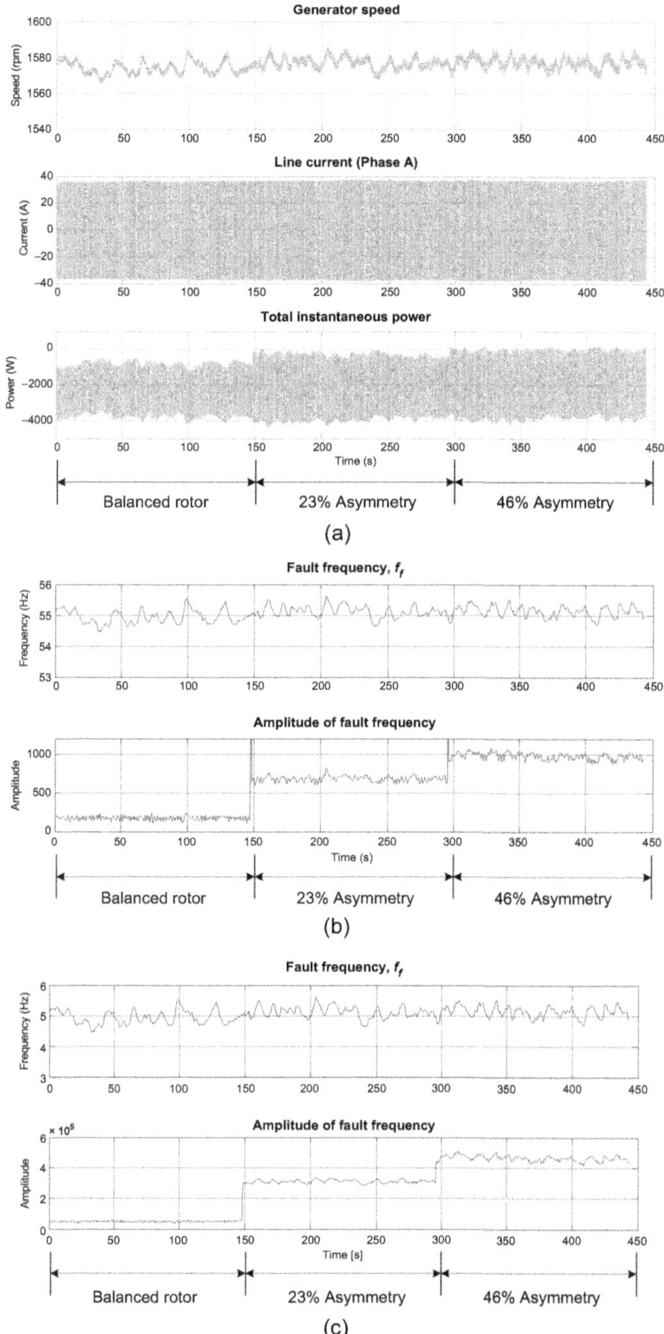

Figure 7.13   *From a variable speed WTCMTR DFIG, spectral component analysis of rotor electrical asymmetry. (a) Electrical signals monitored; (b) line current analysis $(1–2s)f_{se}$; (c) total power analysis sf_{se} [Source: [13]]*

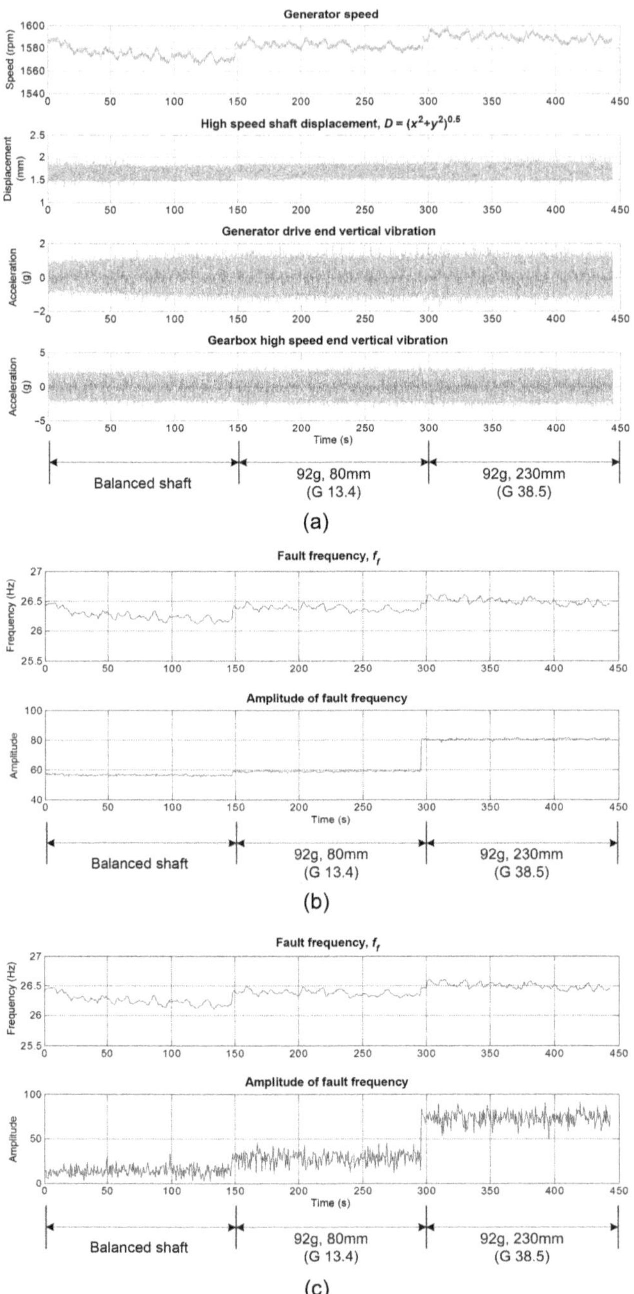

Figure 7.14   From a variable speed WTCMTR DFIG, spectral component analysis of rotor mechanical asymmetry. (a) Raw electrical signals monitored; (b) high-speed shaft displacement analysis f_{rm}, machine rotational speed; (c) narrow band analysis at machine rotational speed, f_{rm}, of gearbox HSS accelerometer [Source: [13]]

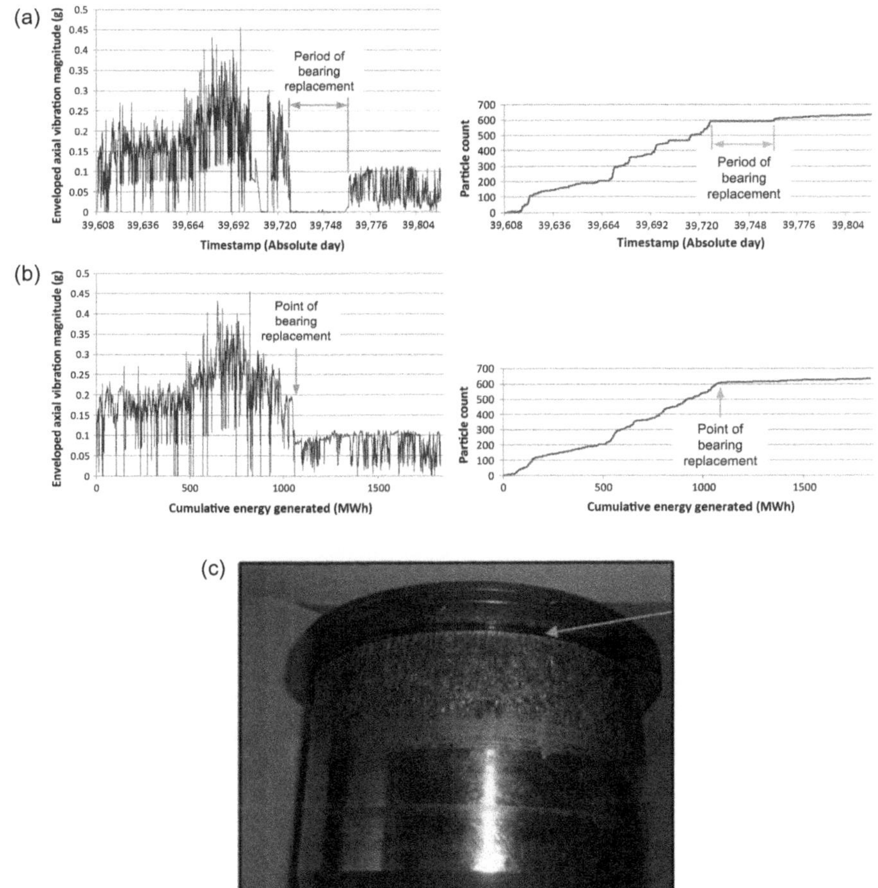

Figure 7.15   *From a ~1.5 MW fixed speed stall-regulated WT gearbox,
detection of incipient IMS bearing failure. (a) HSS axial vibration
amplitude envelope and simultaneous 100–200 μm oil debris
plotted against absolute date stamp; (b) HSS axial vibration
amplitude envelope and simultaneous 100–200 μm oil debris
plotted against cumulative energy generated; (c) IMS bearing
inner race under inspection following its replacement showing
damage [Source: [13]]*

The next example, Figure 7.16, uses narrow band spectral analysis of CMS signals to detect progressive gear tooth failure in the two-stage gearbox of a WTCMTR. The sensitivity of detection shown in Figure 7.16(c) is relatively low but the fault is clearly visible.

The final example of CMS detection, Figure 7.17, looks at the same fault as Figure 7.16 but applied wideband spectral analysis, and the result in Figure 7.17(b) is much more convincing than Figure 7.16(c).

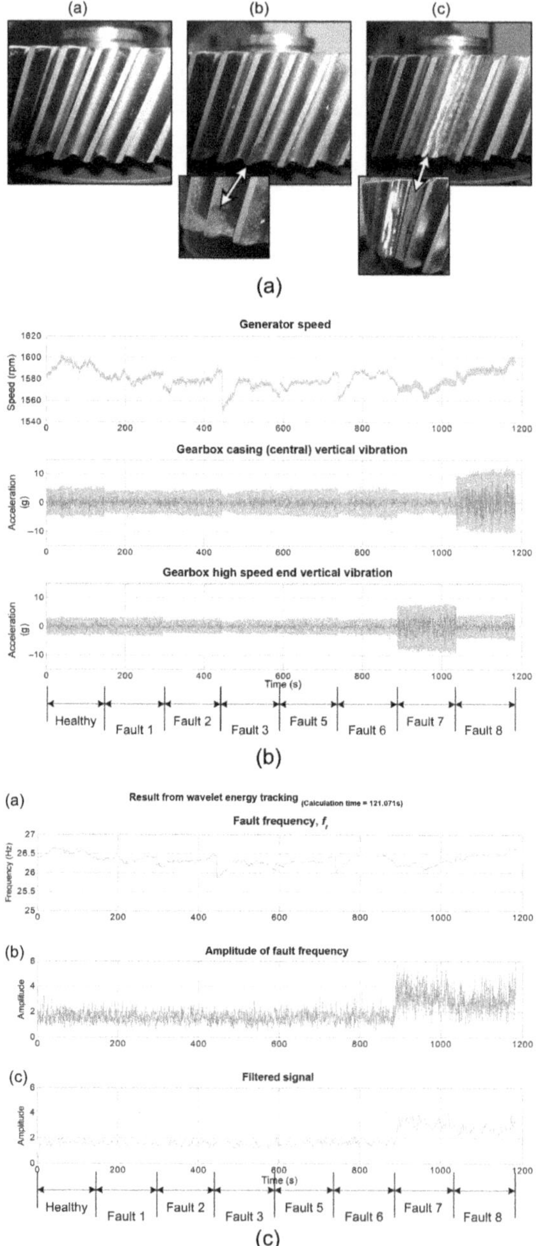

Figure 7.16   From a variable speed WTCMTR two-stage gearbox, detection of gear tooth damage. (a) Progressive gear tooth damage showing fault levels 1, 3 and 8; (b) raw electrical signals monitored; (c) narrow band analysis at machine rotational speed, f_{rm}, of generator DE accelerometer [Source: [13]]

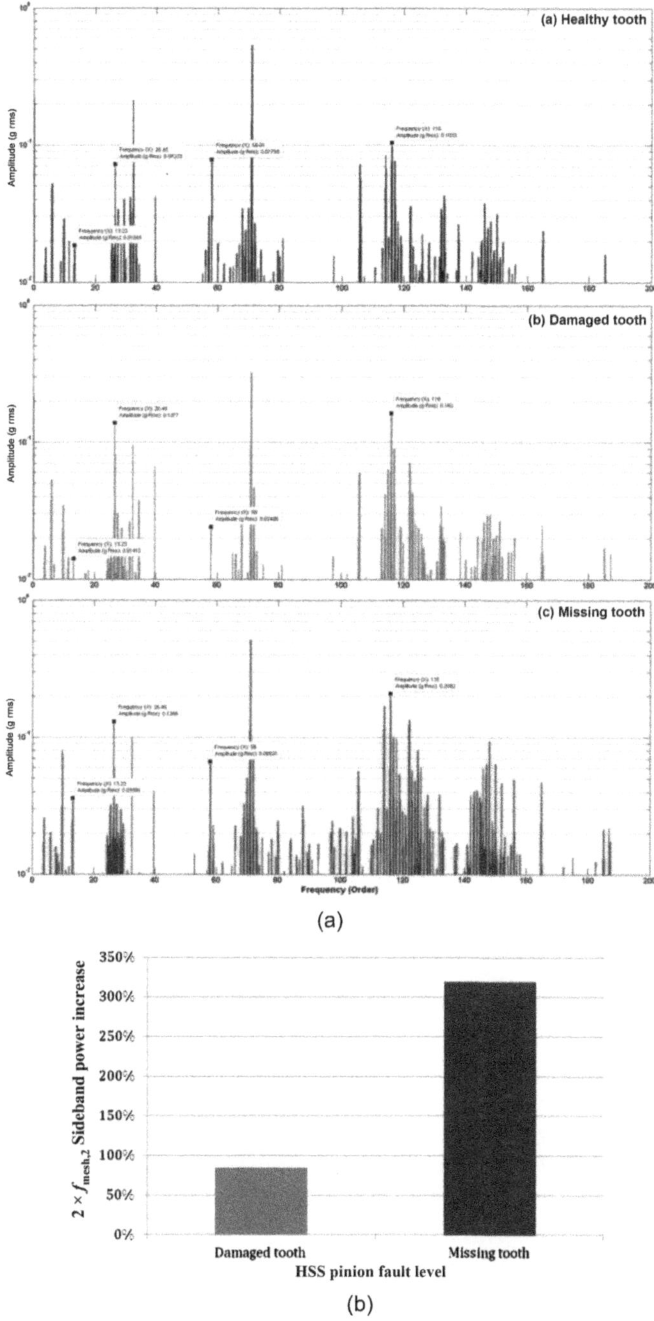

Figure 7.17   From a variable speed WTCMTR two-stage gearbox, detection of same gear tooth damage as Figure 7.16. (a) Gearbox HSS vibration spectra of gear tooth damage; (b) gearbox HSS vibration spectral band amplitudes of gear tooth damage [Source: [13]]

These results show that CMS has a strong potential to detect, diagnose and prognose faults in the WT drive train, particularly in the generator and gearbox.

7.5 Data integration

7.5.1 Multi-parameter monitoring

Despite the potential shown in the previous section, there is clear need for interpretation and the challenge is to achieve that detection, diagnosis and prognosis as automatically as possible to reduce manpower and access costs [13, 14].

One aspect of monitoring that concerns operators is the reliability of the monitoring equipment and the reliability of the detections, exemplified in Sections 7.4.2 and 7.4.3. The former is addressed by experience and the selection of systems from Chapters 13 and 14. The latter depends upon the way the data are presented to the outside world and this can be seen from the figures in this chapter.

There is clear evidence that when a number of monitoring signals from disparate sources present confirmatory fault data, this is helpful and confidence-building to O&M managers and technicians alike. This is clear in the case above of the gearbox bearing fault, as in Figure 7.15, where both vibration and debris count were leading to the same conclusions.

This can be represented as follows, any condition monitoring signal sensor, for example for vibration or temperature, has a probability of detecting a fault in a sub-assembly, for example a gearbox.

The probability of accurate fault detection P_{det} depends in part upon the sensor location P_L and in part upon the sensor reliability P_R.

It has been reported [6] that relying on more than one condition monitoring sensor, for example n sensors or multi-parameter monitoring, almost always increases the chances of successful incipient fault early detection, because if

$$P_{det \cdot n} = 1 - (1 - P_{Rn}P_{Ln})^{n} \tag{7.11}$$

It must be that, provided P_{Rn} and P_{Ln} have reasonable values, that is >50%

$$P_{det \cdot n} > P_{det \cdot 1} = P_{R1}P_{L1} \tag{7.12}$$

That arises simply because of redundancy from sensor failure but more usually because sensors in different locations and of different types raise detection probability, and this must raise confidence in O&M managers and technicians that they are seeing a real effect.

The result of this is that there has been a tendency to swamp machines with sensors, because sensors and data analysis, particularly in SCADA are cheap, which can result in a data overload as we are seeing in the wind industry.

However, there is a law of diminishing returns in (7.11), whilst two sensors may improve the P_{det}, going say from five to six sensors produce a much smaller improvement.

Therefore, operators are encouraged to reduce the number of sensors but raise their quality and compare their signals as a way to higher-quality condition monitoring. This is the rationale for greater integration of the interpretation of signals between SCADA and CMS with the objective of increasing the warning available from detection, which is increasing the prognostic horizon.

7.5.2   System architecture

A barrier to truly integrated monitoring for the WT is the current architecture, exemplified by Figures 7.1, 14.2 and 14.3, where monitoring systems are segregated from one another, largely due to separation between the WT and equipment OEMs. SCADA data, both signals and alarms, are generated within the controller of the WT, whereas the CMS is purchased separately and installed on the WT independently of its controller and it is physically difficult to integrate the CSMS and SCADA signals, notwithstanding their different bandwidths. Some WT controller OEMs, notably Mita Technik [15], are offering SCADA and CMS signal facilities within their controllers, where detection algorithms and alarm handlers can operate on both SCADA and CMS data, comparing trends and extending the prognostic horizon [16]. It is likely that the future of SCADA and CMS integration will lie in this direction.

7.5.3   Energy Technologies Institute project

In the United Kingdom, an important step was taken by the Energy Technologies Institute (ETI) in 2009 to develop a truly integrated WT monitoring system [17]. This project aims to develop a system that can detect the causes of faults and component failures in offshore WTs. It will provide offshore wind operators with sufficient warning to allow a suitable maintenance strategy to be planned, predicting faults before they occur, identifying potential causes and overall, reducing turbine downtime. The system will be planned to have the capability to reduce the *CoE* from offshore WTs.

7.6   Summary

This chapter has described the development of SCADA and CMS in WTs showing that the former is cheap and gives breadth of coverage, whereas the latter is more costly but gives depth of diagnosis.

The chapter has also shown successful examples of SCADA and CMS monitoring giving reliable detection, diagnosis and prognosis of failure modes in the most important sub-assemblies on real WTs in the field and on a WTCMTR.

Good potential for fault detection with significant warning is possible with both monitoring systems, but there is clear evidence that integrating SCADA and CMS data would increase confidence in their indications.

Finally, the chapter suggests how monitoring interpretations can be coordinated to improve information for maintenance planning.

7.7   References

[1]   IEC 61400-25-1:2006. Wind turbines: communications for monitoring and control of wind power plants – overall description of principles and models. International Electrotechnical Commission

[2]   IEC 61400-25-6:2010. Wind turbines: communications for monitoring and control of wind power plants – logical node classes and data classes for condition monitoring. International Electrotechnical Commission

[3]   Germanischer L. *Guideline for the Certification of Condition Monitoring Systems for Wind Turbines, Edition*. Hamburg, Germany: Germanischer Lloyd; 2007

[4]   Watson S.J, Xiang J., Yang W., Tavner P.J., Crabtree C.J. 'Condition monitoring of the power output of wind turbine generators using wavelets'. *IEEE Transactions on Energy Conversion*. 2010;**25**(3):715–21

[5]   Yang W., Tavner P.J., Crabtree C.J, Wilkinson, M. 'Cost effective condition monitoring for wind turbines'. *IEEE Transactions on Industrial Electronics*. 2010;**57**(1):263–71

[6]   Tavner P.J. 'Review of condition monitoring of rotating electrical machines'. *IET Electric Power Applications*. 2008;**2**(4):215–47

[7]   Yang W., Tavner P.J., Wilkinson M.R. 'Condition monitoring and fault diagnosis of a wind turbine synchronous generator drive train'. *IET Renewable Power Generation*. 2009;**3**(1):1–11

[8]   Zaher A., McArthur S.D.J., Infield D.G. 'Online wind turbine fault detection through automated SCADA data analysis'. *Wind Energy*. 2009;**12**(6):574–93

[9]   Gray C.S., Watson S.J. 'Physics of failure approach to wind turbine condition based maintenance'. *Wind Energy*. 2009;**13**(5):395–405. DOI:10.1002/we.36

[10]   Garcia M.C., Sanz-Bobi M.A., del Pico J. 'SIMAP: intelligent system for predictive maintenance application to the health condition monitoring of a wind turbine gearbox'. *Computers in Industry*. 2006;**57**(6):552–68

[11]   Crabtree C.J., Feng Y., Tavner P.J. 'Detecting incipient wind turbine gearbox failure: A signal analysis method for online condition monitoring'. *Proceedings of European Wind Energy Conference, EWEC2010*. Warsaw: European Wind Energy Association; 2010

[12]   Qiu Y., Feng Y., Tavner P.J., Richardson P., Erdos G., Chen B. 'Wind turbine SCADA alarm analysis for improving reliability'. *Wind Energy*. 2012 (in press). Article first published online: 9 DEC 2011 | DOI: 10.1002/we.513

[13]   Crabtree C.J. *Condition Monitoring Techniques for Wind Turbines*. Doctoral thesis. Durham: Durham University; 2011

[14]   Wilkinson M.R. *Condition Monitoring for Offshore Wind Turbines. Doctoral thesis*. Newcastle upon Tyne: Newcastle University; 2008

[15]   Isko V., Mykhaylyshyn V., Moroz I., Ivanchenko O., Rasmussen P. 'Remote wind turbine generator condition monitoring with WP4086 system'. *Proceedings of European Wind Energy Conference, EWEC2010*. Warsaw: European Wind Energy Association; 2010

[16]   Caselitz P., Giebhardt J. 'Fault prediction for offshore wind farm maintenance and repair strategies'. *Proceedings of European Wind Energy Conference, EWEC2003*. Madrid, Spain: European Wind Energy Association; 2003

[17]   Condition Monitoring. Available from http://www.eti.co.uk/technology_programmes/offshore_wind [Accessed 11 December 2011]

Chapter 8
Maintenance for offshore wind turbines

8.1 Staff and training

In Chapter 1, the importance of trained staff for the offshore wind industry was emphasised. This is particularly important in the area of maintenance, most especially offshore, where facilities are restricted and support is limited. WT maintenance technicians require a special blend of technological skills and knowledge, including

- organisation and initiative;
- wind turbine product knowledge;
- mechanical expertise;
- electrical expertise;
- control and software expertise;
- appropriate H&S working practices;
- survival abilities in the offshore environment.

The formation of WT technicians is of great importance to instil appropriate WT product and wind industry knowledge. There has been a split in their training provision between the following:

- Mainstream formation apprentice programmes by WT OEMs in their wind apprentice programmes.
- Specialised wind industry programmes, for example BZEE and from national wind energy associations, suitable for technicians retraining from other fields.
- Specialised H&S and survival training to allow existing onshore technicians to operate safely in the offshore wind farm environment.

It is unlikely that purpose-designed apprentice formation schemes alone will be able to keep pace with the current rate of expansion in the industry, which also needs to attract technicians with relevant basic skills from other industries and retrain them for the wind industry. Trained technicians come into the wind industry from a variety of other relevant maintenance environments, such as power generation, automotive, oil and gas or aerospace and their expertise. Their knowledge is making an important contribution to improving the quality of offshore WT maintenance as well as increasing the quantity of trained staff.

8.2   Maintenance methods

Maintenance methods can be categorised as shown in Figure 8.1. The maintenance strategy for offshore wind is evolving. For onshore wind power the strategy is dominated by preventive maintenance, including planned maintenance scheduled by the WT OEM maintenance manual instructions but affected by unplanned maintenance due to unscheduled stops of the WT.

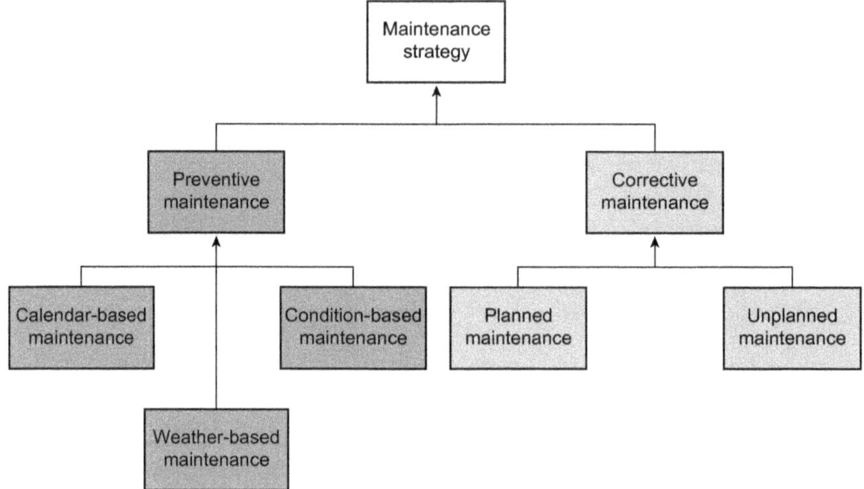

Figure 8.1   Schematic overview of different maintenance types [Source: [1]]

The exigencies of weather offshore and of the difficulties of access in bad weather mean that many preventive maintenance activities cannot be performed when scheduled and need to be done weather and calendar permitting and that induces a degree of planning and preventive action, which is creating a shift in offshore wind O&M management towards a maintenance and asset management strategy.

8.3   Spares

Spares holdings have generally been the responsibility of WT OEMs, but as wind farm sizes have grown, the importance of having key spare sub-assemblies available for rapid change-out has become an important issue and this is of increasing concern offshore where the window for repair, due to weather, logistic and other operational constraints, may be short. Spares holdings probably fall into two categories, major spares with long manufacturing lead times, with a holding that relates to the maintenance and asset management strategy, and consumable spares in frequent and predictable demand, for which the holding can be controlled as consignment stock. These spares can be summarised as follows:

- Major spares:
  - Blades
  - Gearbox
  - Generator
  - Hydraulic power pack
  - Converter inverter modules
  - Pitch motor mechanisms
  - Yaw motor mechanisms

- Consumable spares:
  - Lamps, buttons and control relays
  - Pump motors
  - Filters
  - Grease packs
  - Lubricating oil packs

8.4 Weather

The weather has a major influence on offshore wind farm maintenance, as can be seen from Figures 6.6 and 6.7 and because of the issues presented in Chapter 15, Appendix 6, primarily because of sea state as a result of wind speeds. Availability can go down during the winter season, when wind speeds are generally higher and sea states worsen. This reduction in availability may be partly because of faults caused by worsening weather conditions but is primarily due to the fact that earlier WT faults cannot be repaired because maintenance teams cannot get access to the asset. Figures 6.6 and 6.7 show that availability does not necessarily fall during worsening weather, this means that WT OEMs and operators can avoid these effects with appropriate planning. This means that WT OEMs and operators must plan maintenance during periods of low wind speed when energy resource is not available and access is easy. Weather forecasting has therefore become an important aid to successful maintenance, but forecasting needs to be sufficiently reliable for the WT OEM or operator to give at least 3 days notice of significant weather changes. Currently, WT OEMs and operators use local short-term weather window forecasts for this purpose, but national meteorological offices are developing tools to allow them to tailor their forecasts, accessing national data.

8.5 Access and logistics

8.5.1 *Distance offshore*

The ability to gain good access to the offshore farms is pivotal to achieving the desired reliability and hence availability. This question of access has become more critical with the latest WT wind farm sites that have been awarded in Round 3. To give some idea of the increasing distances involved, Table 8.1 gives the distance

Table 8.1   Rounds 1 and 2 existing wind farms after 2005, >25x WT, distances from shore

Capacity (MW)	No. of WTs	Wind farm name	Min distance offshore (km)	Max distance offshore (km)	Country
90	30	Barrow	7.0		UK
90	25	Burbo Bank	5.2		UK
90	30	Kentish Flats	8.5		UK
60	30	North Hoyle	7.5		UK
60	30	Scroby Sands	3.0		UK
90	27	Inner Dowsing	5.2		UK
97	30	Lynn	5.2		UK
90	25	Rhyl Flats	8.0		UK
90	30	Robin Rigg A	9.5		UK
108	36	Egmond aan Zee	8.0	12.0	NL
120	60	Prinses Amalia	23.0		NL
160	80	Horns Rev	14.0	20.0	DK
165.6	72	Nysted	6.0		DK
110	48	Lilligrund	10.0		SE
		Average	**8.6**	**16.0**	

[Source: [2]]

from land to existing wind farms to date. Table 8.2 summarises the distance from four UK East Coast harbours to two of the largest fields recently awarded by the Crown Estates in Round 3.

It is easy to see from Table 8.1 that physical access was not a major problem for the current wind farm sites 3–23 km offshore. A small vessel will take an hour to reach the furthest field and helicopter flying times will be measured in minutes.

Table 8.2   Distance from major UK East Coast ports to the two largest proposed Round 3 sites

Harbour – Windfarm	Min distance offshore (km)	Max distance offshore (km)
Blyth – Z3 Dogger Bank	118.0	200.6
Blyth – Z4 Hornsea	105.0	212.4
Tyne – Z3 Dogger Bank	112.1	197.1
Tyne – Z4 Hornsea	97.9	206.5
Tees – Z3 Dogger Bank	102.7	194.7
Tees – Z4 Hornsea	76.7	182.9
Humber – Z3 Dogger Bank	107.4	208.9
Humber – Z4 Hornsea	29.5	112.1
Average	**93.7**	**189.4**

[Source: Google Maps and Reference 2]

However, with the new UK Round 3 sites (Table 8.2), 30–212 km offshore, access will become critical. An oilfield support vessel travelling from Blyth would take 10 hours sailing time to the nearest edge of Dogger Bank and 17 hours to the furthest edge of the site.

The advantages and disadvantages of various means of physically getting to offshore wind farms are discussed below.

8.5.2 Vessels without access systems

These vessels (Figure 8.2) have a cruising speed of 20 knots so take ½ to 1 hour to get on site where they then remain on standby till the maintenance crew need to return. They normally have a complement of about 12 personnel and 2 crews. Cooking and toilet facilities onboard make a pleasant working condition for the crews. Their catamaran hull design also makes them very stable. They have been successfully used, again in near-shore wind farms up to 10–20 km offshore in Rounds 1 and 2.

Figure 8.2   Example of an access transfer boat [Source: Alnmaritec]

These vessels tend to rated Marine and Coastguard Agency (MCA) Class 2 allowing them to travel up to 111 km from a safe harbour. However, it is unlikely that they would be used for trips more than 74 km from shore because of the need for a sailing time of 4 hours (Table 8.3). Referring to Table 8.3, it can be seen that these vessels would not be able to cover the Dogger Bank and Hornsea wind farms from the main North East Coast ports likely to be used in UK Round 3.

The perceived advantages of using small vessels would be as follows:

- Simple marine engines that are easily maintained.
- Low cost with fuel consumption at 100 L/hr when cruising at a maximum 30 knots.
- Limited specialist training required for maintenance crews.
- Quick and responsive, already used on sites up to 10–20 km from shore.
- Could be used as an 'infield' vessel launching from a 'mother ship' or fixed platform.

Table 8.3   Calculation of hourly maintenance cost using transfer boats

Hire and fuel costs	Spot market fuel cost £/tonne*	£300
	12 hour trip vessel rental and fuel costs average day rate**	£1500
	Sail out and return journey, 2 × 74 km fuel cost *Based on 20 knot cruising speed giving 0.4 MT fuel used*	£120
	8 hours on location fuel cost *with no heavy seas and light sailing gives 0.4 MT fuel used*	£120
	Total vessel hire and fuel costs/trip	£1740
Hours	Work hours/day estimated at 3 × 4 man crews × 8 hours *This is based on 12 hour shift less 4 hour sailing out and return time*	24 hours
Cost/hour of transfer boat O&M work		**£73/hr**

[Source: *http://www.bunkerworld.com/prices/index/bwi [accessed 5 September 2010]. **http://www.thecrownestate.co.uk/media/211144/guide_to_offshore_windfarm.pdf [accessed 25th May 2012]. Consumption figures from http://www.wildcat-marine.com [accessed 29 June 2010]]

Perceived disadvantages are as follows:

- Weather dependent, especially on sea state, which must be <1.5 m H_s, making the achievement of $>98\%$ accessibility an impossibility.
- Transfer from the vessel to the tower is simple, the boat butts up against ladder and crew members jump onto the WT ladder.
- Only limited amounts of equipment/tools can be transferred from the vessel to the WT.

8.5.3   Vessels with access systems

To achieve the access levels needed to effectively operate an offshore wind farm an oil field support vessel (FSV) is required (Figure 8.3). The size of vessel with dynamic positioning (DP), a computer-controlled system to automatically maintain the position and heading of the vessel using its own propellers and thrusters, and a suitable access system that has been used successfully in the oil and gas industry to access unmanned offshore platforms.

The vessel in Figure 8.3 has a dead weight of 4577 tonnes, is 90 m long with a deck length of 79 m and capable of taking 2500 tonne deck cargo. The crane pictured is heave compensated and capable of lifting 200 tonnes. Heave compensation is a hydro-pneumatic system that takes into account vessel heave to ensure that a crane hook remains stationary relative to the seabed or a fixed object external to the vessel. Maximum draft is 7.8 m. Maximum and cruising speeds are respectively 16.2/12.0 knots and respective fuel consumptions are 62/29 tonnes/day.

Figure 8.3   Example of a field support vessel

The vessel crew is 18 and up to 68 additional personnel can be accommodated if required. It has the ability to stay at sea for 5–7 weeks depending on sea conditions and fuel consumption.

Perceived advantages of an FSV are as follows:

- Achieving required levels of wind farm all access year round.
- Experience in operating these vessels in the oil and gas industry.
- Ability to remain on location to take advantage of short weather windows.
- Capacity to carry a large range of spares and heavier components.
- Enable crews to achieve a longer, more 'stable' shift pattern through facilities on board.

Perceived disadvantages are as follows:

- Potential competition with the oil and gas industry for the same vessels.
- Volatility in the day-rate based on demand.
- Volatility in the fuel bunker price as FSVs consume large amounts of fuel compared with helicopters or small transfer boats described earlier.

The calculation in Table 8.4 shows the hourly cost of maintenance using such a vessel. It assumes four crews working two 12 hour shifts ($2 \times$ day/$2 \times$ night) and covering two WTs at each shift change. With pre-shift briefings and preparation, transfer times and rest periods during the shift, it is estimated 9 hours useful work can be achieved per crew per shift based on experience from the oil and gas industry.

Table 8.4   Calculation of estimated hourly maintenance of an FSV

Hire costs	Average day rate* £10,000.00	
	14 days/trip vessel rental	£140,000
Fuel costs	Spot market fuel £/tonne** £300.00	
	Sail out and return journey (2 × 222 km) fuel cost	£6,948
	Based on 12 knot cruising speed giving 29 tonnes per 24 hours fuel used	
	1 day in port fuel cost	£431
	1.5 tonnes fuel for power generation	
	12 days on location fuel cost	£22,425
	With no heavy seas and light sailing gives 6.5 MT/24 hours fuel used	
	Total vessel hire and fuel costs/trip	**£169,804**
Hours	Work hour/day estimated at 4 shifts × 9 hours (36 hours)	
	This is based on 24 hour working with 2 × day and 2 × night shifts	
	12 days/trip	432 hour
	Cost/hour of FSV O&M work	**£393/hr**

[Sources: All accessed 5 September 2010. *http://www.oilpubs.com/oso/article.asp?v1=9323. **http://www.bunkerworld.com/prices/index/bwi. All consumption figures are calculated from Maersk shipping data]

There are two volatile elements in this costing:

- First is the vessel day-rate, which varies with daily demand and contract duration. The figure used in Table 8.4 is from the oil and gas industry vessel spot market for a 3 month contract.
- Second is the cost of fuel oil, which varies with supply and demand.

Despite these variances, the calculation above does give an indication that the hourly cost compares favourably with helicopters, especially for more distant wind farms.

The main advantage of the FSV, however, is the ability to operate 24 hour working with two 12 hour shifts giving a 8–10 hours useful work on the WT per shift. Such shift patterns are common in the oil and gas industry, so should not be problematic in the wind industry. There are a number of access systems being developed to facilitate the use of these vessels including the Ampelmann, Offshore Access System (OAS), Personnel Transfer System (PTS), Sliding Ladder (SLI-LAD) and the Momac Offshore Transfer System (MOTS).

8.5.4   Helicopters

Although helicopters have been used as a means of transport to and from European and UK Round 1 and 2 wind farms, these have tended to be near-shore, ≤20 km from land, for example Horns Rev (Figure 8.4). The fact that because of visibility requirements the dropping off and recovery of maintenance crews would need to be done in daylight would also limit the time available for WT work, especially in

*Figure 8.4   Example of offshore access to Vestas V80 WTs at Horns Rev by
Eurocopter EC135 [Source: Unifly]*

winter. Psychologically being left offshore without cover, more than 2 hours' flying time away from base may prove difficult for maintenance crews to accept and result in important H&S issues if a casualty occurred. It should be noted that the oil and gas industry has tried to limit helicopters personnel movements because historically and statistically this is the most dangerous aspect of an offshore worker activity.

Another safety consideration is that the fields further offshore cannot be covered by inshore lifeboats and there may therefore be a requirement to have a 'standby vessel' in field to provide safety cover when using helicopters adding to the cost. Such vessels are currently required in the oil and gas industry.

Examples of the cost per hour of maintenance for two types of helicopters are presented in Table 8.5. The figures show that smaller helicopters are cheaper to hire and run. However, it is likely that offshore maintenance crews will not be less than three persons, for safety reasons. Also the need for a helicopter winch operator indicates that the larger helicopters are likely to be necessary for maintenance at distant sites.

Larger helicopters are significantly more expensive due to running/crew costs and are also in demand by the oil and gas industry so the wind industry will be in direct competition for these machines. With the safety briefing, flying and winching time on site, the work period will be very limited, payload for spares/tools will also be limited. In Table 8.5 only two examples of small helicopters have been considered for the following two reasons:

- First is the rotor size, even for a small helicopter with rotor diameter *c.* 10 m, an extension landing basket is required to allow a safe stand off for the helicopter rotor from the WT blades.

- Second, the down-draft generated by the helicopter whilst hovering. In any larger helicopter than the sizes in Table 8.5, the strength of the down-draft may impose unacceptable stress on the WT nacelle and landing basket.

Table 8.6 compares various helicopter rotor sizes and useful payloads, which is proportional to the generated down-draft.

The perceived advantages of using helicopters for offshore WT maintenance are as follows:

- Quick access for assessing maintenance requirements or minor repairs.
- Suitable for close inshore wind farms where the helicopter can be quickly mobilised.
- Fast turn-around for emergency recovery of personnel direct to shore.
- Can operate independent of sea state.

Perceived disadvantages are as follows:

- Helicopter platforms on each WT are expensive, even for large WTs.
- The cost of maintenance operations using helicopters may be prohibitive.

Table 8.5   Calculation of hourly maintenance cost using helicopters

Harbour – Windfarm	Min distance offshore (km)	Max distance offshore (km)
Blyth – Z3 Dogger Bank	118.0	200.6
Blyth – Z4 Hornsea	105.0	212.4
Tyne – Z3 Dogger Bank	112.1	197.1
Tyne – Z4 Hornsea	97.9	206.5
Tees – Z3 Dogger Bank	102.7	194.7
Tees – Z4 Hornsea	76.7	182.9
Humber – Z3 Dogger Bank	107.4	208.9
Humber – Z4 Hornsea	29.5	112.1
Average	**119.9**	**141.5**
Resources	Four-seat helicopter	Pilot + 3 £400/hr
	Seven-seat helicopter	2 Pilots + 5 £1200/hr
Costs	Flight time out (assume from inland ±20 km) assuming Eurocopter EC135, 137 knots cruising speed	1 hour
	Take off, landing, drop off and pick up	0.5 hour
	Flight time in	1 hour
	Total trip	2.5 hours
	Cost of seven-seater flight out £3000 and return £6000	Cost of four-seater flight out £1000 and return £2000
Time	Assuming shift 8 hours–3 hours travel time = 5 working hours	
Cost/hour of helicopter O&M work	**£1200/hr**	**£400/hr**

[Source: http://www.fly-q.co.uk]

- The amount of equipment/spares that can be carried offshore and lowered onto the WT will possibly limit the maintenance to rudimentary servicing.
- They are still weather dependent due to fog/wind/visibility.
- Can only drop off/pick up at the WT in daylight.

Table 8.6   Comparison of various helicopter sizes

	Aircraft	No crew/ passengers	Rotor diameter (m)	Payload (kg)	Range (km)
Small*	Bell 206B-3	1/4	10.16	674	693
	Eurocopter EC135	1/7	10.2	1,455	635
	MBB/Kawasaki BK 117	1/10	11.0	1,623	541
Medium*	Bell 212 Twin Huey	2/13	14.64	2,119	439
	Eurocopter EC155 B1	2/13	12.6	2,301	857
	Sikorsky S-76 Spirit	2/12	13.41	2,129	639
Large*	Bell 214ST	2/16	15.85	3,638	858
	Sikorsky S-92	2/19	17.17	4,990	999
	Eurocopter EC225 Super Puma Mk II+	2/24	16.2	12,633	857
Heavy Lift**	Boeing CH-47 Chinook	3/55	18.3 (×2)	12,495	2252

[Sources: All accessed 9 May 2011. *http://en.wikipedia.org/wiki/Bristow_Helicopters_Fleet. **http://en.wikipedia.org/wiki/Chinook_helicopter#Specifications_.28CH-7D.29]

8.5.5   Fixed installation

Fixed installations are already in use on some offshore wind farms. Their primary use is to house substations and they were constructed using oil and gas platform techniques. To date they have not been continuously manned and are often only used as refuges in the event of rapid change in weather conditions. In the far offshore fields it is highly likely that these installations could be manned all year round or at least for periods such as maintenance campaigns. The substation platform shown in Figure 8.5 is from the Horns Rev 2 wind farm of Denmark.

It is designed as a tubular steel foundation and building. It has a surface area of approximately 20 × 28 m, placed some 14 m above mean sea level. The platform shown as an example accommodates the following technical installations:

- 36 kV switchgear
- 36/150 kV transformer
- 50 kV switchgear
- SCADA, control and instrumentation system and communication unit
- Emergency diesel generator, including 2 × 50 tonnes of fuel
- Sea water-based fire-extinguishing equipment

Figure 8.5   Example of substation installation at Horns Rev 2 [Source: Vattenfall]

- Staff and service facilities
- Helipad
- Crawler crane
- Man overboard boat (MOB)

For more remote fields, the staff and service facilities could easily be upgraded for permanent occupation. The MOB boat could also be upgraded to a transfer boat. The advantage of being on site is that short weather windows could be utilised. Minor WT resets can be quickly achieved and more serious outages quickly investigated, assessed and the information passed back to shore for action.

8.5.6   Mobile jack-up installations

Jack-up installations are mainly used during the construction phase of a wind farm. They give a fixed stable base for cranage to be able to precision lift larger components such as nacelles and blades into position. They also have the advantage of being relatively unaffected by weather conditions once in place with the legs down set on the seabed and the main hull jacked out of the water. They will probably be required during the life of the field for major refits, maintenance or repair jobs that will require large lifting capacity. For more major and longer duration repairs, they provide a fixed platform to work from and can be connected directly to the WT foundation by the means of a gangway that allows for easy continuous access between the workshop facilities on the jack-up and the WT. A prospective jack-up rig is illustrated in Figure 8.6.

Figure 8.6   Example of a mobile jack-up installation [Source: Swire Blue Ocean]

Perceived advantages are as follows:

- Achieves the required level of access year round.
- Experience from operating these vessels in the oil and gas industry.
- Able to remain on location to take advantage of short weather windows.

Table 8.7   Reliability, availability and maintenance data

Item	Data	Data owner	
		In warranty	**After warranty**
1	Baseline reliability data about wind farm components from WT OEMs and other wind farm component suppliers	WT and wind farm component OEMs	WT and wind farm component OEMs
2	WT prototype test data	WT OEM	WT OEM
3	Wind farm component production test data	Operator	Operator
4	Wind farm commissioning data	Operator	Operator
5	SCADA and CMS from WTs and wind farm substation	WT OEM	WT OEM or operator depending on maintenance contract
6	Wind farm maintenance records	Operator/WT OEM	Operator
7	Asset management strategy	Operator	Operator
8	Contractual production targets	Operator/Developer	Operator

- Capacity to carry a large range of spares and heavier components.
- Enable crews achieve a longer more 'stable' shift pattern through facilities on board.
- Provides a stable platform for heavy lifts.

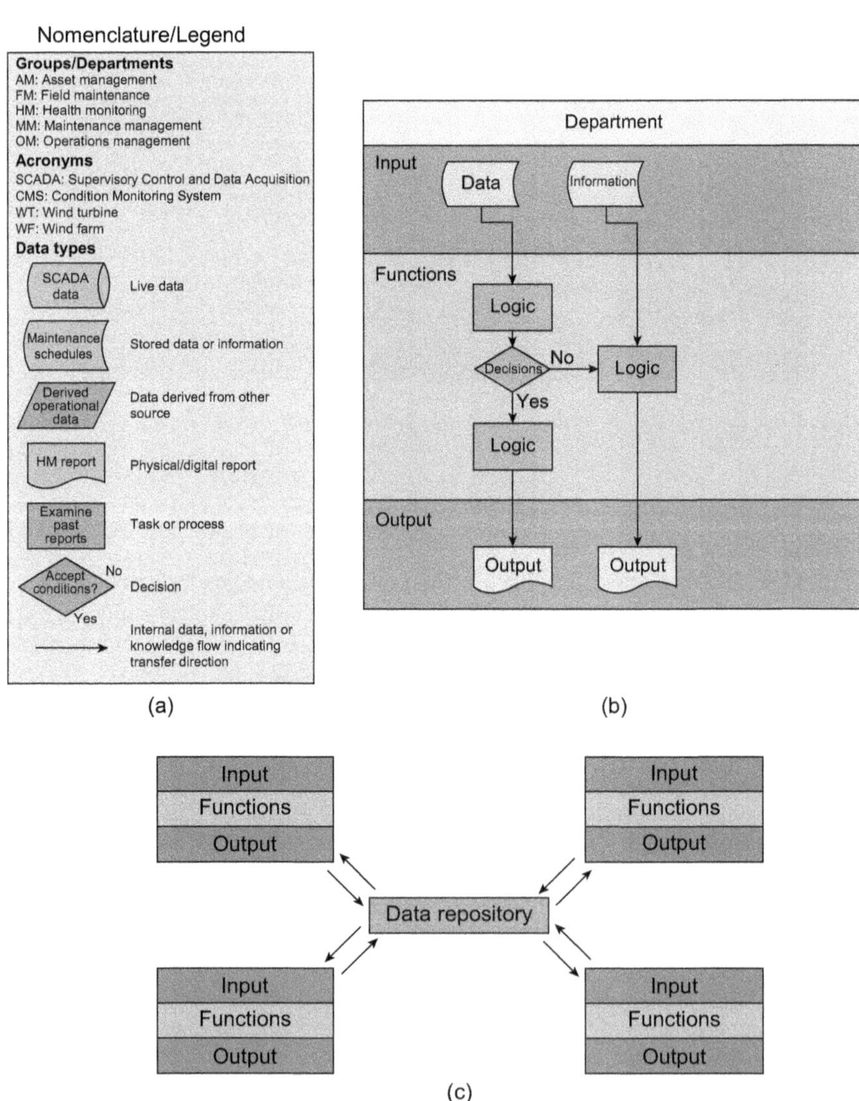

Figure 8.7   Nomenclature, structure and organisation in the proposed Offshore Wind Farm Knowledge Management System. (a) Nomenclature; (b) structure; (c) data flow

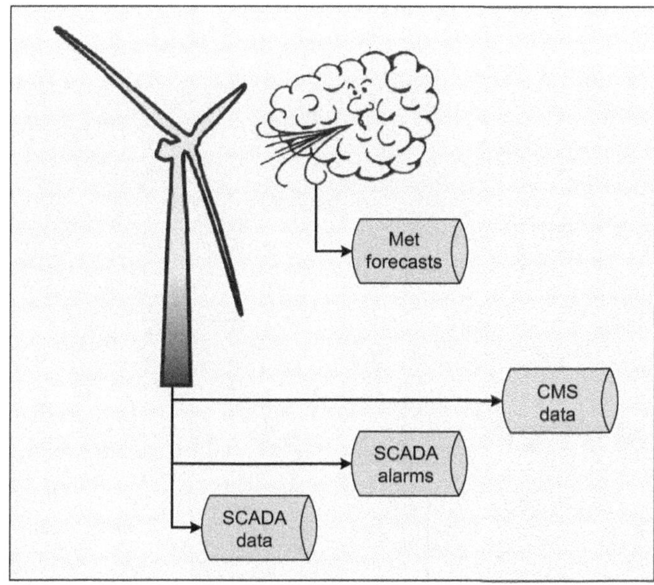

Figure 8.8   Live data produced by an offshore wind farm

Perceived disadvantages are as follows:

- Cost is high.
- Can only operate at one WT at a time.
- Requires good weather to jack-up/jack-down and move between locations.

8.5.7   Access and logistics conclusions

The analysis above shows the following O&M hourly working costs for maintenance logistics: £73/hr, transfer vessel; £393/hr, FSV; £400/hr, small helicopter; £1200/hr, large helicopter. But they also show that whilst access transfer boats are a cost-effective solution for near-shore wind farms, they cannot be effective for the wind farms planned for further offshore [3]. Helicopters have been used successfully for some of those near-shore wind farms (Figure 8.4), but do not have sufficient range and lifting power for the further offshore wind farms where a change would be needed to larger heavy lift aircraft. The alternatives are large FSVs, which can be cost-effective, jack-up vessels, or fixed installations combined with the wind farm substation infrastructure. Each of these alternatives is being tried, but it seems that fixed installations may prove to be the most cost-effective, supported by helicopters and transfer vessels.

An alternative future for distant offshore wind farm accessibility could also be purpose-built vessels. With 20 plus year contracts and wind farms with WT numbers potentially into three figures for a wind farm, financially it will be worthwhile building such vessels at the outset of a new development. These vessels could be

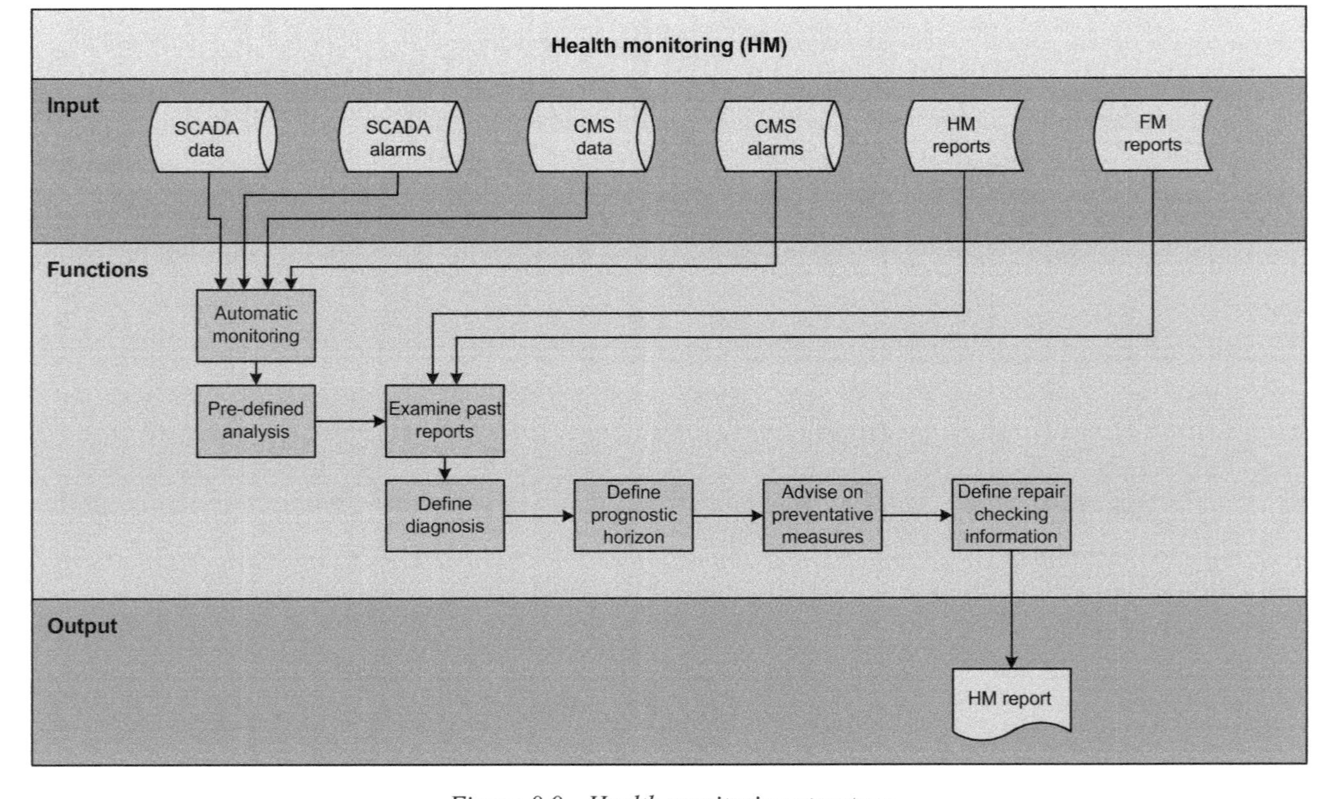

Figure 8.9   Health monitoring structure

Table 8.8 Health monitoring department data

Inputs	Functions	Outputs

Live data
  SCADA signals and alarms data
  CMS signals and alarms data
Stored information
  FM reports
  Health management reports
	i. Apply expert knowledge to monitor WT health via automated SCADA alarm and signal data, CMS alarm and signal data processing
	ii. Examine monitoring results and compare with historical FM reports and HM reports to identify completed repairs, known faults and further deterioration
	iii. Compare observed damage from FM reports with monitoring results to refine diagnostics
	iv. Generate HM reports including fault development, expected time to failure, delaying measures and maintenance recommendations
	v. Define information to allow FM to confirm repair success and include in HM reports
		Reports HM report

semi-submersible or catamaran hull-type design for improved stability and high wave-height operability. They would be dynamically positioned to negate the need to anchor up over seabed cables/utilities and speed-up positioning. Accommodation could be available for up to 100 marine crew technicians and specialists as required. A helideck would allow for crew changes by helicopter or medical evacuation if required. The vessel would be capable of staying on station for some months before returning to port for re-supply. As these vessels have yet to be built, the costs of purchase, hire or operation would be as yet unknown.

8.6 Data management for maintaining offshore assets

8.6.1 Sources and access to data

Data to manage reliability, availability and maintenance come to the operator, maintenance staff and WT OEM from a number of sources. These are set out in Table 8.7. Free access is not available to all this data for the operator because of the contractual arrangements in place. However, it is clear that in order to meet items 7 and 8 in Table 8.7, data from items 4 to 6 should be integrated and measured against a baseline, which may be the performance of other offshore wind farms but should also include measurement against item 1.

The challenge of Table 8.7 is how to integrate that data in an acceptable way to the industry so that it can be worked upon by operator, asset and maintenance management teams and maintenance technicians to meet the strategy and targets of items 7 and 8 and achieve a low cost of energy. This challenge is partly

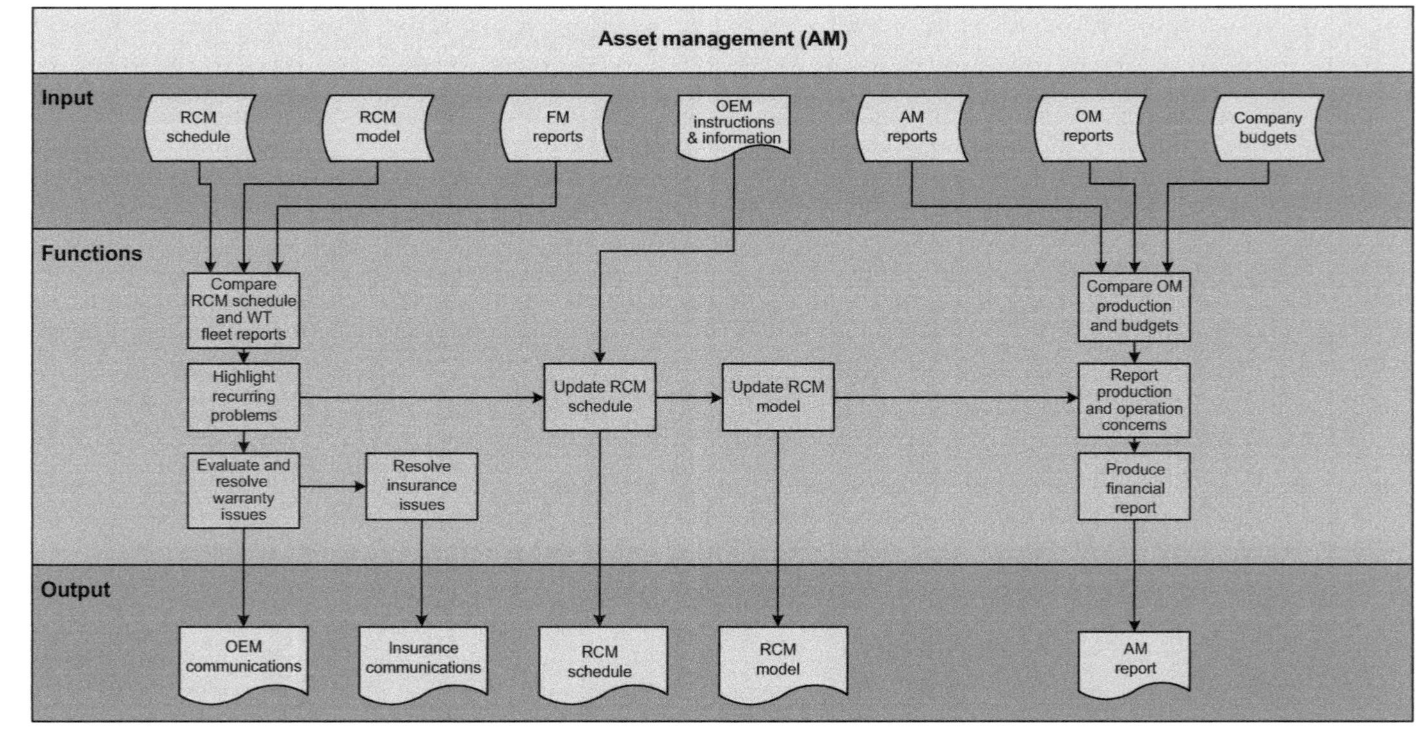

Figure 8.10   Asset management structure

Table 8.9   Asset management data

Inputs	Functions	Outputs

Stored information
    Company budgets
    RCM schedule
    RCM model
    FM reports
    AM reports
    OM reports
External information
    OEM instructions
        and information
	i. Compare OM report outcomes with AM report plans and company budgets and query discrepancies
	ii. Examine reliability from FM reports and compare with RCM model
	iii. Ensure cost-effective use of assets using FM reports, OM reports and past AM reports
	iv. Produce finance reports for inclusion in AM reports
	v. Communicate common and design/type failures with OEM and resolve warranty cases
	vi. Health and safety evaluations (HSE)
	vii. Deal with warranty issues
	viii. Deal with insurance issues
		Reports
		    OEM communications
		    Insurance communications
		    RCM schedule
		    RCM modelAM report

contractual but also technological and needs to feed the design and operation flow charts proposed in Figures 5.4 and 6.8, respectively, in Chapters 5 and 6. The following sections develop such an Offshore Wind Farm Knowledge Management System.

8.6.2   An Offshore Wind Farm Knowledge Management System

8.6.2.1   Structure, data flow and the wind farm

There are a number of closely interlinked industrial groups with involvement in the O&M of an offshore wind farm. The structure described below has been developed within our research group and will need adaptation to the organisation and conditions of individual operators and WT OEMs. The parties can be grouped into six specific departments:

- Health monitoring (HM)
- Asset management (AM)
- Operations management (OM)
- Maintenance management (MM)
- Field maintenance (FM)
- Information management (IM)

The inputs, functions and outputs of each department are defined by means of a block diagram in the format shown in Figure 8.7.

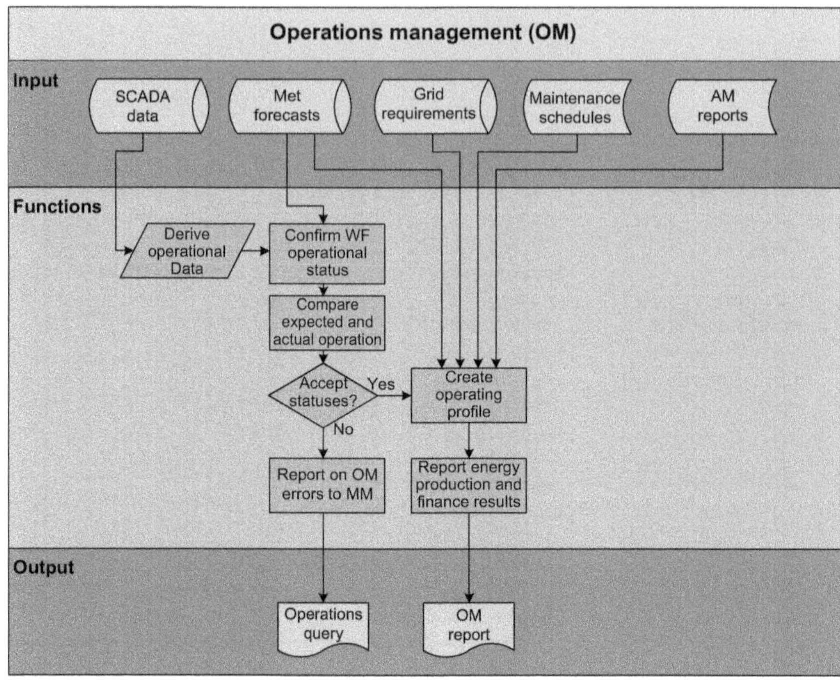

Figure 8.11   Operations management structure

The departments will be laid out in the following sections using the key/ nomenclature and information flows shown in Figure 8.7. A wind farm will also produce various sets of live data shown in Figure 8.8.

Table 8.10   Operations management data

Inputs	Functions	Outputs
Live data SCADA data Met forecasts Grid requirements **Stored information** Maintenance schedules AM reports	i. Compare current WF operating conditions (derived from SCADA) with maintenance reports and query discrepancies with MM ii. Plan and implement WF operating schedules based on grid requirements, Met forecasts, maintenance schedules and AM report iii. Report financial results and energy generated compared to requirements in OM reports	**Reports** O&M reports **Direct reports** Operations query

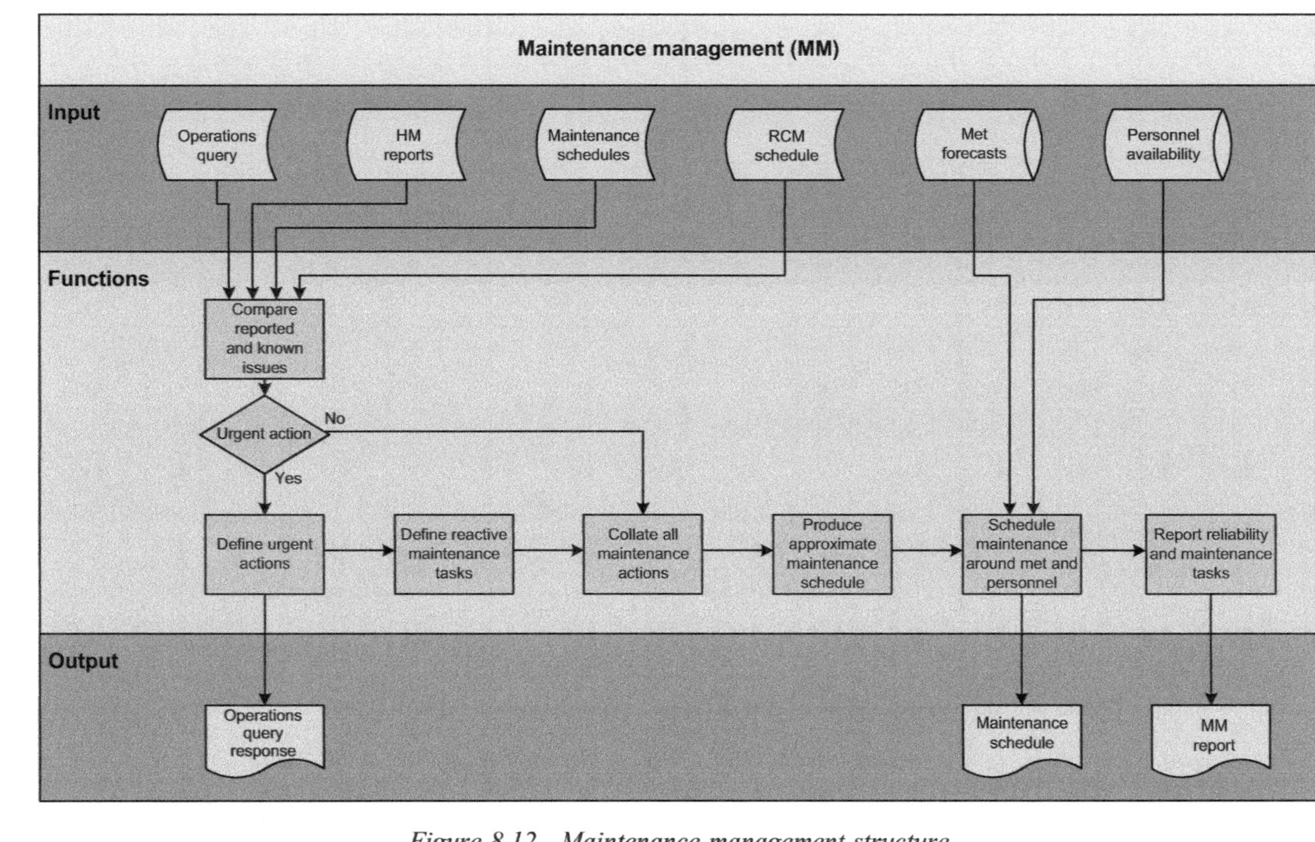

Figure 8.12   Maintenance management structure

8.6.2.2   Health monitoring

Health monitoring (Figure 8.9) is responsible for continuous monitoring of WTs to alert other groups to current and developing faults and advise on their severity (Table 8.8).

8.6.2.3   Asset management

Asset management (Figure 8.10) is concerned with ensuring that operators' assets are operated in the most cost-efficient and valuable manner to secure the longest life cycle of profitable operation (Table 8.9).

8.6.2.4   Operations management

Operations management (Figure 8.11) are concerned with achieving the required wind farm operation, meeting AM and grid requirements (Table 8.10).

8.6.2.5   Maintenance management

Maintenance management (Figure 8.12) are concerned with implementing the requirements of AM, via OM, and responding to concerns raised by HM (Table 8.11).

8.6.2.6   Field maintenance

Field maintenance staff (Figure 8.13) are responsible for the implementation of maintenance schedules and the confirmation of repair success (Table 8.12).

Table 8.11   Maintenance management data

Inputs	Functions	Outputs

Live data
    Met forecasts
    Personnel
        availability
Stored information
    Maintenance
        schedules
    HM reports
    RCM schedule
Direct reports
    Operations query
	i. Compare operations queries and HM reports with maintenance schedules (known issues)
	ii. Respond to operations queries
	iii. Compare issues with RCM schedule
	iv. Detail maintenance tasks including preventative, reactive and RCM responses
	v. Produce approximate cost-effective maintenance schedule
	vi. Produce updated maintenance schedule based on met forecasts and personnel availability
	vii. Ensure RCM activities meet AM plans
	viii. Report initial reliability figures in MM report
		Reports
		    Maintenance
		        schedule
		    MM report
		    Operations query
		        responses

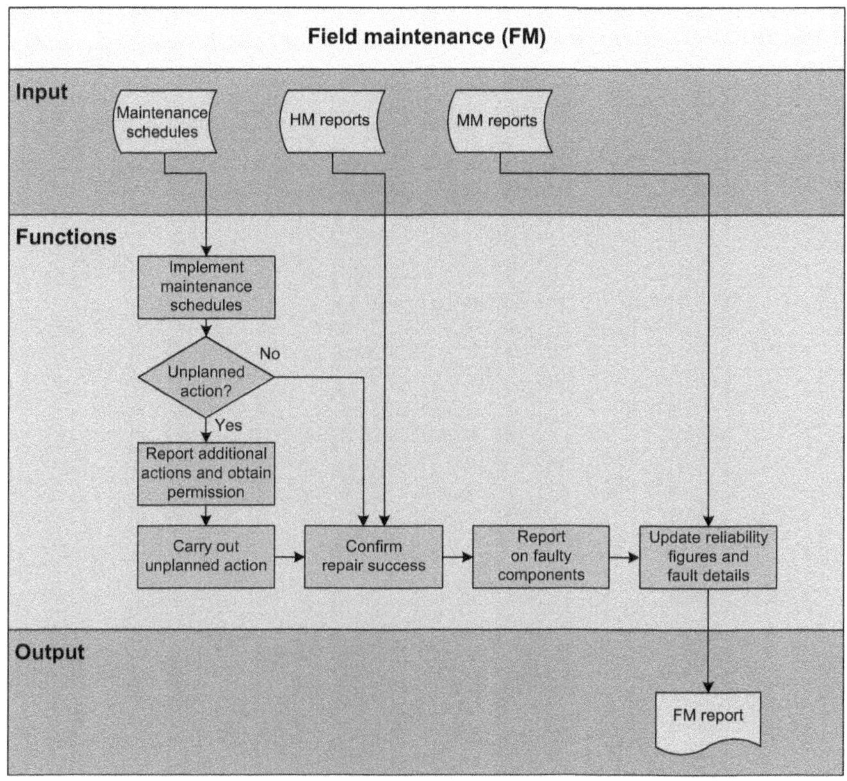

Figure 8.13   Field maintenance management structure

Table 8.12   Field management data

Inputs	Functions	Outputs
Stored information MM reports Maintenance schedules Health management reports	i. Implement maintenance schedules ii. Report and resolve any faults or potential faults discovered during maintenance iii. Confirm repair success against advice in HM reports iv. Update reliability figures and fault details from MM report and insert into FM report v. Report actions taken and results of faulty component examination in FM report	**Reports** Field management report

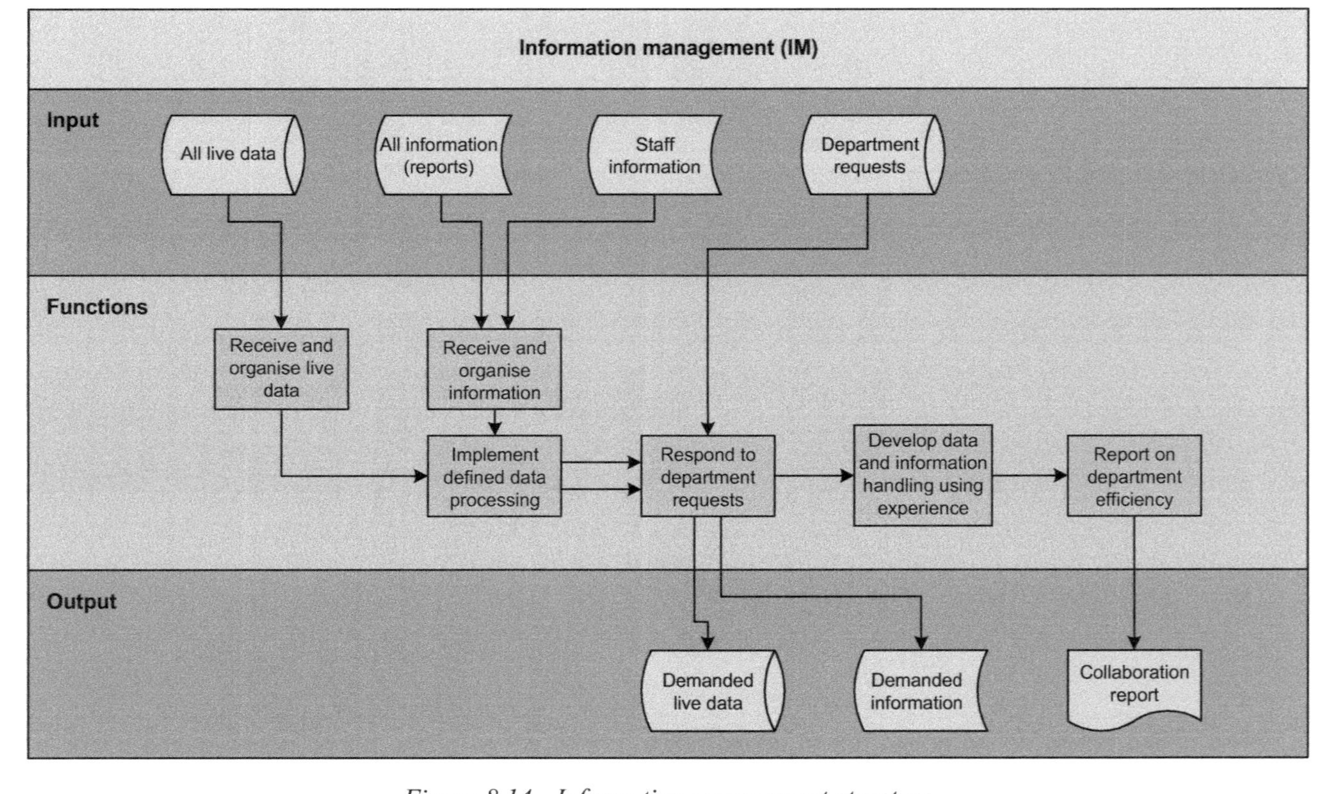

Figure 8.14   Information management structure

Table 8.13   Information management data

Inputs	Functions	Outputs
All reports All live data Staff information Department requests	i. Receive **live data** and **information** from other departments ii. Manage and store reports in central repository iii. Alert departments to new reports iv. Provide **on demand data and information** based on **department requests** v. Provide information analysis support to all departments vi. Perform database maintenance and updates vii. Control the removal of data for database maintenance viii. Realise effective communication between departments and report this efficiency and strategy in a **collaboration report**	Collaboration report On demand data and information

8.6.2.7   Information management

Information management (Figure 8.14) handles the data produced by the wind farm (Table 8.13).

8.6.3   Complete system

Finally, the complete structure for this data is given in Figure 8.15 and represents a proposal for an Offshore Wind Farm Knowledge Management System.

8.7   Summary: towards an integrated maintenance strategy

This chapter has presented the personnel, infrastructure and data issues associated with improving the maintenance, availability and reliability of offshore wind farms. The key factors are shown to be the training of staff, the availability of appropriate access infrastructure and the presentation of appropriate data from the wind farms to those staff to allow that costly infrastructure to be fully exploited.

A proposal has been presented for an Offshore Wind Farm Knowledge Management System to handle the SCADA and CMS monitoring data and the accumulated reliability data of the wind farm correlating it with the maintenance logs to provide an integrated system upon which maintenance planning can proceed.

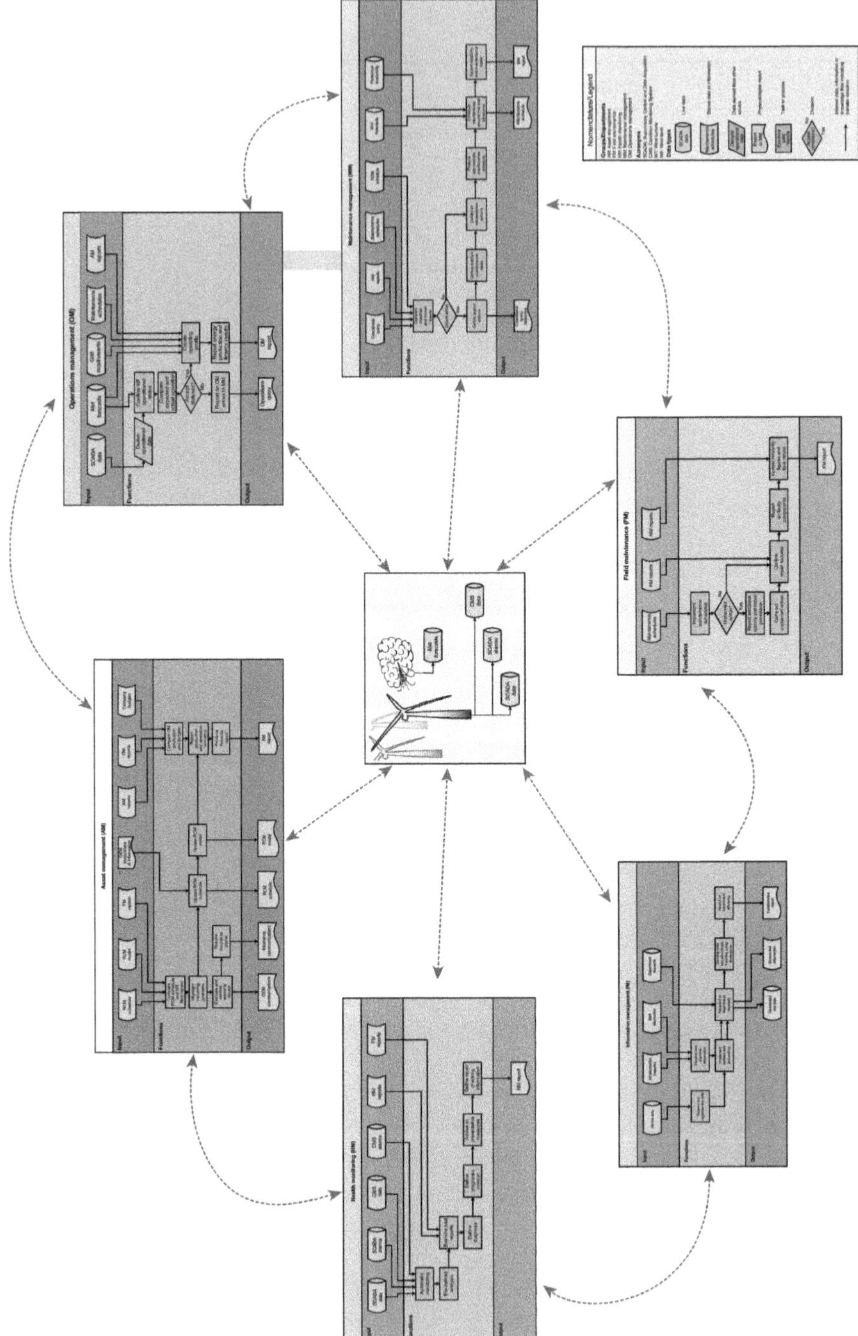

Figure 8.15   *Large offshore wind farms: data management monitoring and maintenance*

8.8   References

[1]   Wiggelinkhuizen E., Verbruggen T., Braam H., Rademakers L., Xiang J., Watson S. 'Assessment of condition monitoring techniques for offshore wind farms'. *Journal of Solar Energy Engineering.* 2008;**130**(3):1004-1–9

[2]   Richardson P. *Relating Onshore Wind Turbine Reliability to Offshore Application.* Master of Science Dissertation, Durham University, Durham; 2010

[3]   Bierbooms W.A.A.M., van Bussel G.J.W. 'The impact of different means of transport on the operation and maintenance strategy for offshore wind farms'. *Proceedings of European Wind Energy Conference, EWEC2003.* Madrid, Spain: European Wind Energy Association; 2003

Chapter 9

Conclusions

9.1 Collating data

From the preceding chapters it would seem that the keys to higher availability and lower cost of energy for offshore wind farms will be

- metrics of the availability and reliability expected of the wind farm and its component WTs;
- a clear maintenance strategy to achieve those metrics;
- a clear asset management strategy to support the maintenance strategy through time to the full asset life cycle.

The diagrams in Figure 5.4 followed by Figure 6.8 show the sequence of tasks needed to develop reliable WTs and highly available wind farms. Figure 6.8 concentrates on their operation and both figures show the importance of data to provide the metrics to drive these processes. Maintenance strategies to be deployed in offshore wind are summarised in Figure 9.1.

Onshore WT maintenance has been typified by corrective maintenance on the right hand side of Figure 9.1. The result can be seen in Figure 9.2, taken from Windstats onshore WT survey data [1]. Maintenance is being equally time distributed amongst sub-assemblies (Figure 9.2(b)) with no regard to the downtime consequences of sub-assembly failures (Figure 9.2(a)). So, for example, the gearbox causes 22% of the downtime but receives only 8% of the maintenance time, similar to the hydraulics that caused only 6% of the failure downtime. Whilst this approach may have been acceptable onshore, where time absorbed was facilitated solely by maintenance technicians accessing the WT by a low-cost van, it will clearly not be acceptable offshore where every maintenance visit incurs sea- or airborne access costs, described in Chapter 8.

9.2 Operational planning for maintenance, RCM or CBM

Reliability-centred maintenance (RCM) is where WT sub-assembly failure rate and downtime are used to drive maintenance activity. Therefore, from Figure 9.2, maintenance time on the gearbox would be arranged to be 22% of total, bearing in mind the downtime the gearbox causes. This distribution of maintenance will vary with time, depending on the performance of WTs and their sub-assemblies.

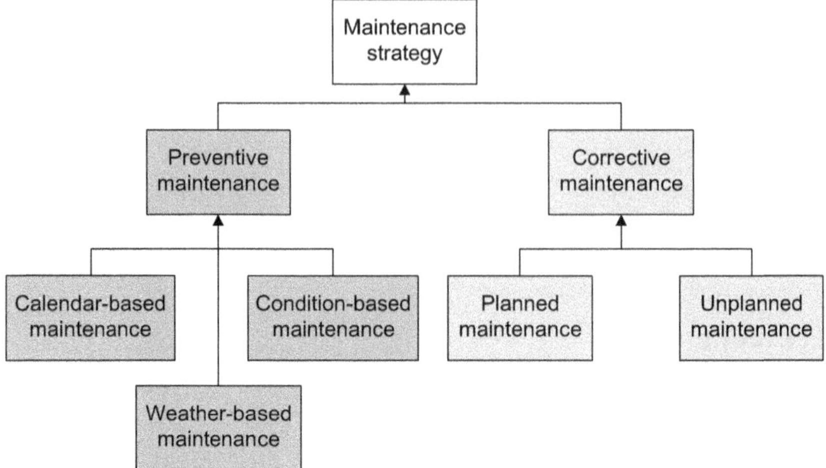

*Figure 9.1   Schematic overview of different maintenance strategies highlighting
onshore on right, offshore on left*

However, such an approach may be misguided unless the maintenance activity genuinely reduces failure sub-assembly rates and downtimes. How can this be determined? It can only be determined by having a clear understanding of sub-assembly history and performance. This can be achieved from the reports described in Chapter 8, that is, RCM.

Or, it can be achieved by monitoring the performance of the WT using methods like those described in Chapter 7, that is, condition-based maintenance (CBM). WTs have exceptionally good monitoring cover because of their unmanned remote robotic operation, but very few operators are making use of the monitoring information to manage their maintenance because of the volume and complexity of the data. That must change offshore. The data must be simplified and presented in a coordinated, comprehensible way, hence the need for a data management system. It must then be used to drive RCM and CBM to raise availability and lower cost of energy. Both RCM and CBM drive the need for an Offshore Wind Farm Knowledge Management System.

9.3   Asset management

RCM and CBM address the ongoing operation of the wind farm but they cannot, on their own, secure the through-life reliable performance of the wind farm without longer-term management of the asset [2]. The high capital cost of offshore wind demands a rigorous operational regime that generates energy at an adequate price that recovers the cost of the investment. But once payback is achieved, the life of the asset will determine its long-term profitability. These longer-term benefits can only be secured by long-term management of the asset, that is, controlling the later part of the bathtub curve (Figure 5.5) where wear out of sub-assemblies is controlled by their planned change out. It seems clear that offshore wind farms, with

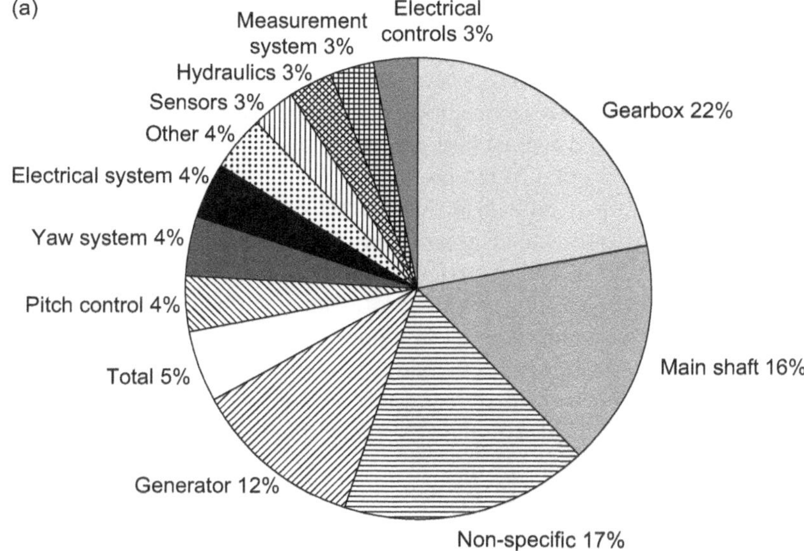

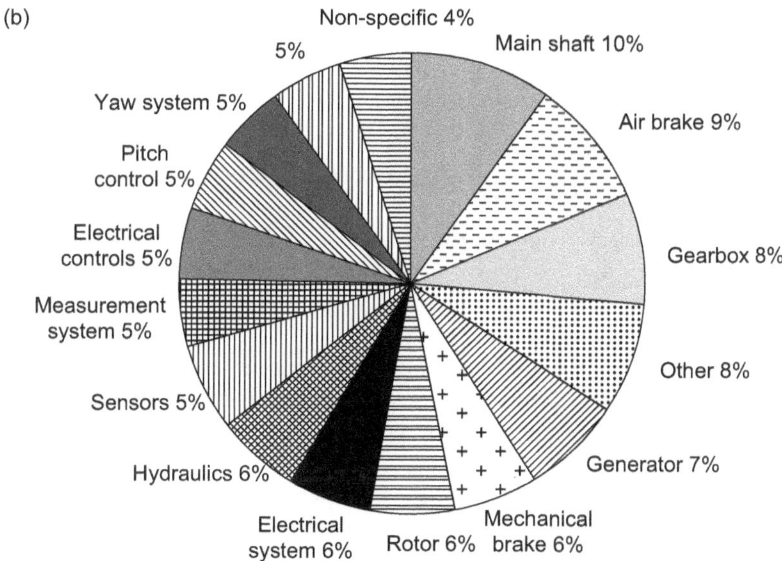

*Figure 9.2   Comparison of downtime to maintenance time per sub-assembly.
(a) Downtime per sub-assembly; (b) maintenance time per sub-assembly*

large numbers of identical or similar WT assets, can benefit from planned change out of the most vulnerable sub-assemblies: blades, gearboxes, generators, converters and even nacelles. In fact, that change out process can also embed sub-assemblies with improved operational and reliability performance.

9.4   Reliability and availability in wind farm design

The author suggests that enormous assistance in the above task would be rendered by the ready availability of more reliability data from OEMs and operators to allow designs to be benchmarked against best practice. This is the kernel of the reliability proposals in Chapters 5 and 6 of the book.

In the early days, the wind industry was secretive about performance, to protect its intellectual property (IP) and champion individual improvement. Operators have also been protective of wind farm performance data as it has contractual value.

But the industry is now of a size and professionalism where it must find a way to share data within the wind industry in a non-competitive way to champion collective wind power improvement as the industry comes into direct competition with fossil- and nuclear-fired and other renewable power sources. The wind industry must share data if it is to deal with the CAPEX and OPEX challenges offshore and meet the competition head-on [3].

An important reliability and availability issue, in terms of cost, will be to determine maintenance cost-effectiveness. Some operators are setting availability targets for offshore wind farms. There may be dangers in this approach, since higher availability can always be achieved with higher O&M investment. The better path will be to determine the optimal O&M costs to achieve an acceptable availability and that will vary from wind farm to wind farm, depending strongly on the location of the site, being affected by distance offshore, local infrastructure and assets and their costs.

9.5   Prospective costs of energy for offshore wind

What has become clear from writing this book is that the high capital cost of offshore wind means that much greater attention is now being paid to making the asset work at a high availability to achieve its projected payback targets than the wind industry has been accustomed to onshore.

This does not mean that this cannot be done, since it is already being achieved at Baltic offshore wind farms, with availabilities of 96–98%, and a growing number of North Sea and Irish Sea wind farms moving towards higher availability, 90%–95%.

This means that investors, developers and operators are looking more critically at the intrinsic reliability of offshore wind farm and its components than was ever considered onshore.

9.6   Certification, safety and production

The design of WTs is regulated by a certification process that ensures the strength and safe operation of the WT designed. Furthermore, the WT control systems are designed to ensure this safe operation. Stiesdal and Hauge-Madsen [4] said 'the classical principle of wind turbine control and monitoring is to ensure that the wind turbine is always in a safe state – this is not automatically the same as ensuring that the operating time is maximised'.

Offshore wind during installation and operation is also a potentially hazardous activity. Developers and operators have therefore rightly adopted a strong certification- and H&S-oriented approach to the development of new WTs and to their installation and operation. Many H&S lessons have been learnt as staff were trained, and this approach must be maintained as the industry moves to deeper and more distant waters and new staff are drawn into the industry. The H&S lessons learnt already mean that near shore the offshore wind industry can start to combine a certification and H&S-oriented approach with a more production-oriented approach as happened in the North Sea oil and gas production in the 1990s.

As the wind industry matures, the current certification- and H&S-oriented approach is likely to change, as the more stringent demands for return on the larger capital outlays for capital projects encourage a more vigorous production-oriented approach. In this stage of development of the industry the interaction between operators, asset managers, certifiers, insurers and investors will need to be strengthened and attention will shift to more attention being paid to O&M issues and through-life costs.

It could also well be that in 10 years we will see the onshore wind industry learning from the more structured operational environment of the offshore wind industry.

9.7 Future prospects

The future prospects for offshore wind power look favourable. Early wind farms have demonstrated that high levels of resource are available but at considerable CAPEX and OPEX costs, emphasising the importance of

- reducing initial CAPEX costs;
- designing high-reliability wind farm assets to reduce prospective risk;
- reducing the cost of and risk of asset deployment;
- managing O&M to restrain OPEX costs;
- within that framework, achieving low cost of energy figures by achieving as high WT availability as is practical for the location of the wind farm.

Experience with onshore wind has shown that though the initial capital costs are high, the distributed nature and repeatability of the technology are such that the learning curve time constant, probably 5 years for onshore, is short and that many manufacture, installation, operation and maintenance lessons are rapidly learnt. The offshore wind learning curve time constant is clearly longer but probably of the order of 7–10 years judging from Figure 1.6.

9.8 References

[1] *Windstats (WSD & WSDK) quarterly newsletter, part of WindPower Weekly, Denmark.* Available from http://www.windstats.com [Last accessed 8 February 2010]

[2]   Bertling L., Allan R.N., Eriksson R.A. 'Reliability-centred asset maintenance method for assessing the impact of maintenance in power distribution systems'. *IEEE Transactions on Power System.* 2005;**20**(1):75–82

[3]   Barberis Negra N., Holmstrøm O., Bak-Jensen B., Sorensen P. 'Aspects of relevance in offshore wind farm reliability assessment'. *IEEE Transactions on Energy Conversion.* 2007;**22**(1):159

[4]   Stiesdal H., Hauge-Madsen P. 'Design for reliability'. *Proceedings of European Wind Energy Offshore Conference EWEAO2005*; Copenhagen, 2005

Chapter 10

Appendix 1: Historical evolution of wind turbines

Year	Development	Associated technology	Photo
200 BC	Wind machines used in Persia		
AD 70	Hero's Pneumatica – reaction steam turbine	Debate exists whether Hero invented it or was stimulated by other examples to make one	
Seventh century AD (Wiki) 1000s (Shepherd 1998) by the Rashidun Caliph Umar AD 634–644	First practical windmills were built in Sistan, Iran, Persia–Afghanistan border region of Sistan, for grinding grain and pumping water, 50 of them were in operation until 1963 in Neh, Iran	VAWT, vertical axle, long vertical-driven shafts, rectangle blades Enclosed by a two-storey circular wall, millstones at the top, rotor at the bottom Rotor: spoked with 6–12 upright ribs, each covered with cloth to form separate sails [Hau]	
1119	The Netherlands	Post-mill, HAWT Functions: draining water, milling grain, sawing wood. Easy to yaw, but support might be an issue, from post-mill to cap-mill, the background	

(Continues)

(*Continued*)

Year	Development	Associated technology	Photo
1191	First windmill in England in West Suffolk		
1219	Chinese VAWT	Sheng Ruozi quotes a written selection about windmills from the 'Placid Retired Scholar', actually Yelü Chucai (1190–1244), a prominent Jin and Yuan statesman, after the fall of Jin in 1234 to the Mongols. The passage refers to Yelü's journey to Turkestan, in modern Xinjiang in 1219, and Hechong Fu is actually Samarkand in modern Uzbekistan	Adjustable or luffable sails, that is, self-adjusting sail direction in response to wind condition as the windmill is rotated. Ancient Chinese windmills (B Zhang, 2009)
1200s	Squat structure, wooden shutters, in Europe	4 blade, HAWT	Shutters are adjustable, that is, luffable blades.
1295	The Netherlands	HAWT cap- or tower-mill. Post-mills dominated the milling and pumping scene in Europe until the nineteenth century when tower-mills began to replace them The advantage of the tower-mill over the earlier post-mill is that it is not necessary to turn the whole mill body or buck, with all its machinery into the wind; this allows more space for the machinery as well as for storage	

(*Continued*)

Year	Development	Associated technology	Photo
Early 1500s	The Netherlands 'Wipmolen' hollow post-mill	Driving scoop wheels for pumping water	The Wipmolen was a more compact tower-mill, which could be described as a cap-mill where the yawing machinery was concentrated in the cap of the mill
1800s	The Netherlands	Development of precision wooden pin and socket gears for cap-mills	Patent for wooden right-angle gearbox between the horizontal mill rotor axis and the vertical wallower gear axis Horizontal to vertical rotary power Check the year
1854	Daniel Halladay formed the Wind Engine and Pump Company where it became one of the most successful windmill companies, Batavia, Illinois, USA	HAWT, multi-blade up-WTs for water pumping Automatic yawing Innovation: design and manufacturing excellence Availability of steel facilitated this rotor technology	
1866	Pumping water on farms, filling railroad tanks, USA	HAWT, multi-blade upwind turbines. Application of US mass-production methods to large, remote, mechanised farms	

(*Continues*)

(*Continued*)

Year	Development	Associated technology	Photo
1883	First automatically operated windmill for electricity production for battery charging by Charles Brush, in Cleveland, Ohio, USA	12 kW, HAWT, 144 blade WT Innovation: combined available US WT manufacturing techniques with new electrical generation methods DC generator had only been available for 5 years in USA and Europe, prior to the diesel and petrol engine	
1887	Prof. James Blyth of the Royal College of Science, Glasgow, now known as Strathclyde University, for electricity production for battery charging	10 kW, VAWT, 4-blade WT driving a DC generator; believed to have had some adjustable or luffable blades, which contemporary alternatives did not	
1887	Poul la Cour, Denmark, for electricity production for battery charging	10 kW, HAWT, 4-blade, fixed-pitch WT driving a DC generator Innovative aerodynamic system	
1888–1900	Experimental windmills were used to generate electricity in USA and Denmark, based on designs of Halladay and Poul la Cour	The need for electricity for pumping and light, on large, remote, mechanised farms, in the flat, windy mid-west US, stimulated US wind power development	
1900–1910	Many electric windmill plants were in use in Denmark, 2500 windmills up to 30 MW in total	Flat, windy Danish landscape. Did Danish immigrants to USA contribute la Cour technology to USA?	

(*Continued*)

Year	Development	Associated technology	Photo
1908	72 electric WT generators recorded in Denmark	5–25 kW, HAWT, D 23 m, 24 m high, 4 blade	
1910–1930	USA produces 100,000 farm HAWT windmills/year for water pumping	Mixture of the American and Danish designs Proof of high quality of US mass-production techniques and the need for power where there was a lack of grid connection	
1910–1914	Diesel engine competition for electric windmills	Following the development of the diesel- and then petrol-engine-driven generators	
1914–1918	First World War, reduction in oil supplies, 20–35 kW electric windmills were built		
1918 Post-war	Windmill development languished	Small WTs were proving less reliable for electricity production than diesel- or petrol-engine-driven generators Also grid connection was becoming more widespread	
1930s	Windmills for electricity were common on large farms in Denmark and USA	High-tensile steel was cheap, and windmills were being placed atop pre-fabricated open steel lattice towers	The beginning of decline of the American multi-blade turbine concept
1920s		Influence of aerodynamic knowledge from aircraft following the First World War, for example, development of the wing and propeller This started to affect WT design	

(Continues)

(*Continued*)

Year	Development	Associated technology	Photo
1931	In Yalta, USSR, modern WT	100 kW, 30 m high, HAWT, geared drive, 3 blade, connected to 6.3 kV distribution system; 32% capacity factor; adjustable blade flaps Post-mill with the whole structure rotating along a track Early large 3-blade machine exhibiting clear signs of growing aeronautical influence	
1938–1944	Denmark F.L. Smidt	45 kW range, 2 blade, a significant number installed annually in Denmark	
1939–1945	Second World War, another reduction in oil supplies, increases wind power development		
1940	Ventimotor company formed with a test centre near Weimar, Germany, to develop WTs for the German war effort and included Ulrich Hutter among its key personnel	Excellent aerodynamics, light and cost-effective	

(*Continued*)

Year	Development	Associated technology	Photo
1941	In USA, operation of Smith–Putnam, the world's first megawatt-size WT connected to the local electrical distribution system on Grandpa's Knob, Castleton, Vermont, USA, designed by Palmer Cosslett Putnam and manufactured by the S. Morgan Smith Company, perhaps the grandfather of the modern electrical WT	1.25 MW, D 57 m, 40 m high, HAWT, 2-blade, geared drive, constant speed, full-span pitch control, stall-regulated, downwind turbine. Sophisticated modern WT. First grid-connected WT	
1945 Post-war	National electrification of Europe and North America using fossil-fired power stations. Research programmes considered wind power as a supplement in Denmark, France, Germany and UK		
1945–1970	New growth in wind power took place, principally in Western Europe and particularly in Denmark under the direction of those trained by Poul la Cour		

(*Continues*)

(*Continued*)

Year	Development	Associated technology	Photo
1954	Costa Head in Orkney. First experimental grid-connected WT in UK by John Brown Engineering Company	100 kW, D 18 m, HAWT, 3-blade, geared-drive, pitch-regulated, downwind turbine; slip ring induction generator grid connected; lack of marketing, demand or mass production	
1956–1966	Station d'Etude de l'Energie du Vent at Nogent-le-Roi in France operated an experimental WT	800 kVA, HAWT, 3-blade, geared-drive, pitch-regulated, downwind turbine, interesting design but no subsequent development, probably because of French national decision to concentrate on nuclear power	
1950s	100 kW, D 25 m, downwind, 2-blade, pneumatically driven generator	Enfield–Andreau turbine at St. Albans, UK	
1956	In Denmark, Juul developed at Gedser the modern WT, forerunner of the Danish Concept and considered to be the mother of the modern electrical WT	200 kW, D 24 m, HAWT, geared-drive, 3-blade, stall-regulated, upwind turbine, with aerodynamic tip brakes on rotor blades, released automatically in over-speed. Blade tip brakes are a good innovation	

(*Continued*)

Year	Development	Associated technology	Photo
1972	International oil crisis triggered by the Yom Kippur War and a renaissance of wind power		
1976–1981	Modern small-scale WTs	1–10 kW VAWT and HAWT	Small-scale inheritors of the 1930s US and Danish small turbines. The market was still uncertain because the technology was still unresolved
1979	Carmarthen Bay, UK, VAWT 450	130 kW, VAWT with furling blades, very unusual and did not work	
1979	In Denmark, at Nibe, two experimental machines were erected, one with pitch control and one without	200 kW, D 24 m, HAWT, geared-drive, 3-blade, fixed-speed, stall-regulated, upwind turbine with aerodynamic tip brakes on rotor blades	

(*Continues*)

(*Continued*)

Year	Development	Associated technology	Photo
	If Gedser was the mother of the modern WT, these were her two strongest children	630 kW, D 40 m, HAWT, geared-drive, 3-blade, full-span pitch control, fixed-speed, stall-regulated, upwind turbine	
1980s	MBB, $30 million 1-blade Monopteros WT programme; three 600 kW prototypes still in service near Wilhelmshaven; programme featured a line of WTs from D 15–56 m	600 kW, D 15–56 m, HAWT, geared-drive, 1-blade, upwind turbine Very novel, high performance, light weight Some are still operating, but the concept is not popular with customers	
1980s	Great California wind rush; large numbers of WTs ≤ 100 kW, mostly HAWT but some VAWT	Very poor reliability of many designs	
1980	In the Netherlands, development of a modern WT	300 kW, geared drive, 3-blade, stall-regulated, fixed-speed WT	

(*Continued*)

Year	Development	Associated technology	Photo
	In USA, MOD 0	200 kW HAWT, geared drive, 2-blade, full-span pitch control, downwind turbine	
	In USA, MOD 1	2 MW HAWT, geared drive, 2-blade, full-span pitch control, downwind turbine Overweight and unreliable	
1981	In USA, Boeing, MOD 2	2.5 MW, D 91 m, HAWT, geared drive, 2-blade, full-span pitch control, upwind turbine Sophisticated light weight; but no teeter hub so excess stress at centre of blade	
1982	In Sweden, WTS 75-3	2 MW, HAWT, geared drive, 3-blade, full-span pitch control, upwind turbine	
1982	In USA, WTS4	4 MW, HAWT, geared drive, 3-blade, full-span pitch control, downwind turbine Sophisticated design Huge and complex	

(*Continues*)

(*Continued*)

Year	Development	Associated technology	Photo
1983	In Germany, Growian, Große Windenergieanlage, Germany, invested $55 million in this WT, which operated for only 420 hours before experiencing fatigue failure in the hub	3 MW, D 100 m, 100 m tall, HAWT, geared drive, 2-blade, full-span pitch control, downwind turbine, with fully rated cycloconverter Very unusual, big and risky, unreliable	
1987 or 1988	Prototype large grid-connected WT designed and constructed by Wind Energy Group, at Burgar Hill, Orkney UK	3 MW, D 60 m, HAWT, geared drive, 2-blade, full-span pitch control, upwind turbine Very unusual, big, risky and unreliable	
1987	At Richborough in UK a large grid-connected WT	1 MW, HAWT, geared drive, 3-blade, stall-regulated, fixed-speed, upwind turbine, with on rotor aerodynamic tip brakes Simple, rugged, reliable, but lack of market penetration	
2002	In Germany, Enercon E-112	4.5–6 MW, D 112 m, HAWT, direct drive, 3-blade, full-span pitch control, upwind turbine Son of Growian but with good reliability.	

(*Continued*)

Year	Development	Associated technology	Photo
		All electric with fully rated converter connected to the grid. The owner of Enercon, Alois Wobben, is a power electronic engineer	
2010	Norway, Statoil's Hywind project	Siemens SWT2.3, 3-blade, upwind, geared drive, variable speed, pitch-regulated turbine mounted on a floating, moored caisson	

Chapter 11

Appendix 2: Reliability data collection for the wind industry

11.1 Introduction

11.1.1 Background

WT manufacturers, operators, maintainers and investors agree that it is essential for WTs to have a high reliability to achieve a high capacity factor and availability and thereby deliver electricity at a low cost of energy. An important factor in achieving those objectives is that WTs, when designed, should have the highest possible reliability. Currently, the European wind industry is achieving WT availabilities onshore of 96–97% and offshore of 90–95%. It would be desirable to raise these availabilities, and design for reliability would contribute to that aim.

An important requirement of design for reliability is to be able to measure, predict and analyse WT reliability using accurately defined mean time to failure (MTTF), mean time to repair (MTTR) and mean time between failures (MTBF) data for WTs. These standard terms are defined by International Standards and are listed in Section 1.6.1.

The definition of the terminology and taxonomy of wind turbines and the collection of reliability data and its interrelationship with WT design, defined by IEC 61400 [1], needs to be standardised. It is also clear that in order to increase WT reliability, more and higher quality reliability data is needed from the wind industry, within limits of commercial confidentiality.

This appendix is a proposal from the EU FP7 ReliaWind Consortium for the standardisation of

- taxonomy of the wind turbine,
- English terminology for the naming of components,
- methods for collecting reliability data from wind turbines in the field,
- methods for reporting failures from wind turbines in the field.

The purpose of these standardisations is to improve wind turbine reliability in the field, to raise wind turbine availability and to lower the consequent cost of energy. These issues also affect other industries, including offshore oil and gas, power generation, transportation, military and aerospace. An example of reliability data collected from the first of these industries, oil and gas, is shown by OREDA [2]. A standard for the collection of reliability data from that industry also exists, EN ISO 14224:2006 [3].

11.1.2   Previously developed methods for the wind industry

The most detailed previous public domain WT data collection campaign was funded from 1996 to 2006 by the German Federal Ministry for Economics & Technology under the 250 MW Wind Test Programme, which included the Wissenschaftliche Mess- und Evaluierungsprogramm (WMEP), Scientific Measurement and Evaluation Programme, administered now by Fraunhofer IWES Institute [4]. This was built on an earlier work by Schmid and Klein [5]. A standard failure report form was used by WT operators for return to IWES. This form is given in Appendix 3. Schmid and Klein [5] have also given valuable examples of data collection forms. The proposals below are drawn from this experience.

11.2   Standardising wind turbine taxonomy

11.2.1   Introduction

This section summarises the general principles and guidelines on which the taxonomy will be based, and the taxonomy is derived from a deliverable prepared for the EU FP7 ReliaWind Consortium by the author and other consortium members.

The taxonomy should be adaptable for application to the common reliability analyses needed for WTs, such as failure mode, effects and criticality analysis [6], failure rate Pareto analysis, reliability growth analysis and Weibull analysis.

The intention of adopting such a taxonomy would be to overcome current deficiencies of the data collection, which can be summarised as follows:

- consistency of naming of the systems, sub-systems, assemblies, sub-assemblies and components of WTs;
- non-traceability of the system monitored;
- unspecified WT technology or concept;
- problems of confidentiality between parties when exchanging data.

11.2.2   Taxonomy guidelines

A WT taxonomy is a structure that names the main features of a WT in a standard terminology exemplified in Figure 11.1.

- The taxonomy must be reliability oriented, particularly in respect of analysis. It is agreed that such an approach is the best compromise between the various needs of an industry, which leads to a different system breakdown, grouping and terminology than would be achieved by the description of simple components.

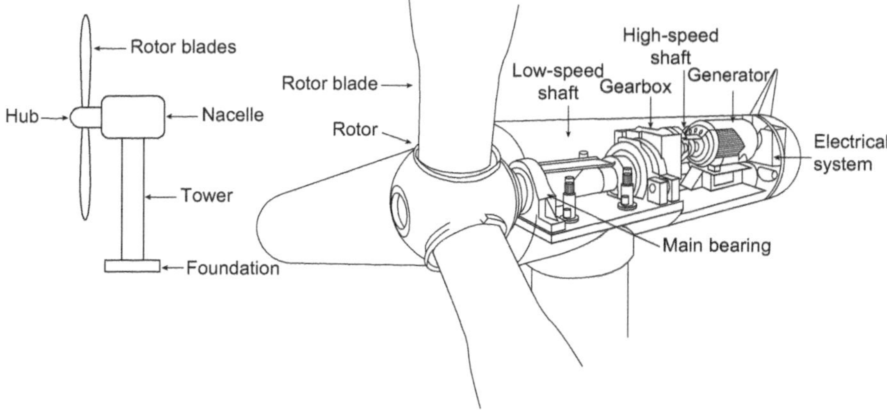

Figure 11.1   Example of a WT and its nacelle layout showing the terminology

- The taxonomy will include all the WT concepts' components in five levels. Data will be retrieved on the basis of a concept code, which allows WT model mapping within the taxonomy for any given data set.
- The taxonomy is based on a Danish concept WT, which is an upwind, three-bladed, horizontal axis, un-ducted WT. Other concepts could be included upon achievement of significant industrial uptake by this taxonomy.
- At the highest level, outside the taxonomy, the WT concept should be identified by a code. For example, indicating stall-, active stall- or pitch-regulated, fixed or variable speed, geared- or direct-drive, doubly fed induction, induction or squirrel cage induction or wound or permanent-magnet synchronous generator. Therefore each item in the taxonomy will be clearly linked to a code associated with each WT concept.
- The taxonomy should also inform the structure of the monitoring input/output (I/O) applied to the WT, whether that is for signal condition and data acquisition (SCADA) or condition monitoring system (CMS) signals and alarms because the taxonomy will be used to focus on SCADA and service log data available from operational wind farms. Therefore the terminology of components in the I/O list of the SCADA [7] should agree with the component names used in the taxonomy.
- The taxonomy shall be organised in five indented levels. Each level should be justified with a brief description that shall include the rationale for the level grouping and intended use.
- The first five indented levels of the taxonomy must comply with the Figure 11.2 using Table 11.2 as an example. The taxonomy may not reach the lower level components, for example to individual electronic capacitors, but an analyst could add additional lower levels, if needed, but they must be compatible with the upper five levels of the taxonomy. Analysts could also add additional

elements in levels 1–5, if absolutely necessary for their purpose, denoted by the prefix CUSTOM, although this customization should strongly be discouraged.

- The taxonomy will have a short code alpha-numeric designation for each item at each level. It is anticipated that the construction of the designation could follow the guidance of Reference 8, which adopts an alpha-numeric code although some in the wind industry prefer a word code.
- The lowest level components will be grouped according to the following two concepts:
  - Functional grouping for the signalling, supervisory and control components, examples: pitch encoder grouped with a control and communication system, LV electrical systems grouped together.
  - Positional grouping for mechanical components, examples: gearbox, pitch system, blade, frequency converter, generator, blade.

For example: generator temperature sensor and pitch encoder are both components of the monitoring system. This segregation is necessary due to the nature of WT systems that signalling, supervisory and control components tend to spread throughout the WT, whereas mechanical devices are located in a specific position within the WT. This is exemplified in Table 11.1.

- In case of ambiguity, the designation will follow the order mentioned above: first the functional groupings, second the positional groupings.
- At the lowest indented level the component name should have no ambiguity with similar components of different assemblies. For example: the pitch pinions and the yaw pinions.

11.2.3   Taxonomy structure

The structure of system, sub-system, assembly, sub-assembly and component that should be adopted is shown in Figure 11.2. The WT itself is considered as the system.

Examples of this terminology are shown in Table 11.2.

A full taxonomy, listing sub-systems, assemblies and sub-assemblies, is provided in Section 11.6.

Table 11.1   Examples of parts groupings

Functional grouping	**Positional grouping**
Control and communication system	Generator
Lightning protection system	Pitch system
110 V Electrical auxiliary system	Gearbox
220 V Electrical auxiliary system	Yaw system
400 V Electrical auxiliary system	Blade
WT power system	Hub
SCADA system	Main shaft set
Collection system	Foundation
Grid connection	Tower
Hydraulics system	

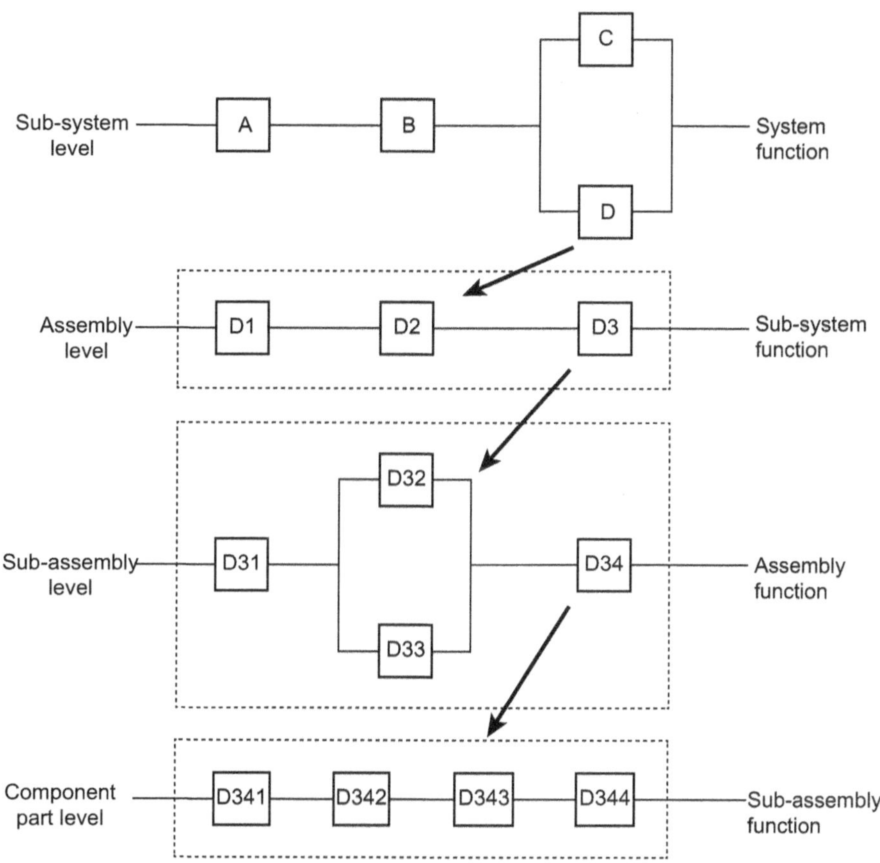

Figure 11.2   Example of system, sub-system, assembly, sub-assembly and component structure, cf. Figure 2.7

Table 11.2   Examples of application of terminology

System	Sub-system	Assembly	Sub-assembly	Component
Wind turbine	Rotor	Electrical pitch system	Pitch motor	Brush
Wind turbine	Drive train module	Gearbox	Gearbox	Stage 1 planetary wheel
Wind turbine	Electrical module	Frequency converter	Power electronics	IGBT

11.3   Standardising methods for collecting WT reliability data

In the ReliaWind Consortium the following method was used where it was proposed that reliability data from WTs should be collected in five tables as follows (Table 11.3).

Table 11.3   Events

Wind farm	Turbine ID	Date and time of event	Time to repair TTR (hours)	Actual repair time ART (hours)	System	Sub-system	Assembly	Sub-assembly	Component	Failure mode	Root cause	Maintenance category	Severity category	Additional information
A	1	2008-04-01 11:28:01	54.2	N/A	Wind turbine	Drive train	Gearbox assembly	Gearbox	N/A	N/A	N/A	4	3	N/A
A	23	2008-04-24 01:56:11	168.4	3.5	Wind turbine	Rotor	Pitch system	N/A	N/A	N/A	N/A	3	2	N/A
B	2	2008-04-25 08:43:24	2.5	1	Wind turbine	Power	Generator assembly	Generator	Stator phase b winding	Open	Over current	1	1	Series failure
…	…	…	…	…	…	…	…	…	…	…	…	…	…	…

The fields are defined as follows:

Wind farm	*Confidentiality will require that this be an anonymous identifier*
Turbine ID	*Turbine identifier within the wind farm*
Time of event	*Time stamp in ISO form yyyy-mm-dd hh:mm:ss*
Mean downtime (MDT)	*Total number of hours during which the turbine was not operational i.e. includes all the time needed to restore the WT to an operating condition*
Time to repair (TTR)	*Actual number of hours completing the repair i.e. excludes logistics associated with the repair action such as having the component delivered to site, arranging technicians' time*

System structure	*The system structure used is that set out in Section 11.2.3*
Sub-system	*Select from an approved list, see Section 11.6; it should usually be possible to ascribe a failure to a particular sub-system*
Assembly	*Select from an approved list, see Section 11.6; it should usually be possible to ascribe a failure to a particular assembly*
Sub-assembly	*Select from an approved list, see Section 11.6; it may not always be possible to ascribe a failure to a particular component*
Part	*Select from an approved list; it may not always be possible to ascribe a failure to a particular component*
Failure mode	*The particular way in the failure occurred, independent of the reason for failure; this may be subjective, but very useful if available*
Root cause	*The cause of failure may be subjective, but very useful if available*
Maintenance category	*A description of the maintenance impact of the failure:* *1. Manual restart* *2. Minor repair* *3. Major repair* *4. Major replacement*
Severity category	*A description of the severity of the failure based on MIL-STD-1629A Section 4.4.3, which relates to the ability of the system to carry out the function for which it was defined safely and efficiently:* *1. Minor* *2. Marginal* *3. Critical* *4. Catastrophic*
Additional information	*Pertinent comments where available*

The list of events in Table 11.3 will be exhaustive within the following criteria:

- The event required manual intervention to restart the machine.
- The event resulted in downtime ≥ 1 hour.
- There will be no missing events or missing time periods; or if there are, the missing time periods will be noted and the reasons stated.
- Every cell in the table should have either a data value or be filled with N/A, not available.

Table 11.4 is derived entirely from Table 11.3 and no new information is added. It is thought unlikely that enough details will be available in Table 11.3 to permit the calculation of failure rate on a per component basis. Failure rates should be reported per year as standard but information from Table 11.6 will allow calculation per operational period in a year, per GWh in a year, per revolution, or some other metric, depending on what information is available for the particular wind farm. Confidentiality may require that this information be aggregated on a wind farm, rather than WT basis. This information could be presented graphically, for example as shown in Figure 3.6(a).

Table 11.5 is also derived entirely from Table 11.4 and no new information is added. The downtime should be given in units of hours. Confidentiality may require that this information be aggregated on a wind farm, rather than WT basis. This information could be presented graphically, for example as shown in Figure 3.6(b).

Table 11.4   Failure rates

Wind farm	Turbine	Sub-system	Assembly	Year			
				1	**2**	**3**	**...**
A	1	Drive train	Gearbox	0	2	1	...
A	1	Power	Generator	2	1	2	...
A	1	Rotor	Pitch	1	2	1	...
...	...		...	...	...	...	...

Table 11.5   Downtime

Wind farm	Turbine	Sub-system	Assembly	Year			
				1	**2**	**3**	**...**
A	1	Drive train	Gearbox	24	5	1	...
A	1	Power	Generator	65	4	2	...
A	1	Rotor	Pitch	21	5	5	...
...	...		...	...	...	...	...

Table 11.6   Wind farm configuration

Wind farm	Turbines	Rated power (MW)	Mean wind speed (m/s)	Mean turbulence intensity	Hub height (m)	Rotor diameter (m)	Terrain type	Control type	...
A	20–40	1–2	6–8	0.25–0.50	60	40	Offshore	A	...
B	0–20	2–3	8–10	0.50–0.75	55	30	Onshore exposed	B	...
...	...	...	...	...	...	...	...	...	...

Table 11.7   Additional turbine information

Wind farm	Month	Energy generated (GWh)	Revolutions	...
A	2008-01	50	1.544×10^5	...
A	2008-02	70	2.422×10^5	...
...	...	...	...	...

There would be two versions of Table 11.6:

- Table 11.6(a) will have all values stated exactly and to maintain End User confidentiality will remain private; and
- Table 11.6(b) will be available to a consortium but will be less specific about machine characteristics, with identifiable parameters categorised into appropriate ranges to make anonymous the data as shown in the example above.

The control type column will be populated from a standard list. Further columns may be added to this table depending on what information is available for each wind farm.

Confidentiality requirements may mean that the information in Table 11.6 could not be publicly available. For a wind farm to be included in the survey it is desirable that the site contains at least 15 turbines that have been running for at least 2 years since commissioning. Data for the tables above should be provided by WT operators.

11.4   Standardising downtime event recording

The approach recommended is to describe and classify downtime events as stoppages of duration ≥ 1 hour and requiring at least a manual restart, categorising downtime events as follows:

- Category 1: manual restart
- Category 2: minor repair
- Category 3: major repair
- Category 4: major replacement

11.5   Standardising failure event recording

11.5.1   Failure terminology

When a failure has occurred it is important to record the details of that failure. In the WMEP Failure Report Form given in Appendix 3 a simple tick box approach was adopted.

This provides insufficient detail for maintenance and root cause analysis purposes, and the following approach is suggested for recording failures in detailed fault or maintenance logs, taken from the recommendations in Reference 4. The terminology to be used should be that shown in Section 11.6, which is consistent with the proposed Structure shown in Figure 11.2 using Table 11.2 as an example. The failure modes suggested there include those identified by WP partners of ReliaWind WP2.

11.5.2   Failure recording

This section provides a broad method of failure recording, rather than trying to capture every different possible failure mode. For example a bearing failure could encompass:

- Inner race failures
- Outer race failures
- Cage failures
- Element failures

The recommended failure recording terminology is in part recursive, referring successively to the component, sub-assembly, assembly, sub-system, system, shown in Figure 11.2 using Table 11.2 as an example, in turn. For example, using Section 11.7 for recording a gearbox epicyclic bearing failure, the failure description should take the following format:

- Bearing failure: planet bearing: epicyclic part: gearbox: drive train: wind turbine.

11.5.3   Failure location

Location indicators are needed for components, such as bearings, where several may be found in a single assembly or sub-assembly, in that case the following rules could be followed, using the failure example above:

- If more than one epicyclic stage exists in a gearbox, the first stage is that closest to the WT rotor and so on.
- In a parallel shaft gear train, the pinion drives and the gear are driven.
- The two ends of a gearbox are the rotor end or generator end.
- Where there are two bearings on a gearbox shaft, the one closer to the gear should be referred to as the inner bearing and that further from the gear as the outer bearing.
- Generator bearings should be defined as drive end (DE) and non-drive end (NDE).

11.6 Detailed wind turbine taxonomy

System	Sub-system	Assembly	Sub-assembly	Component
Wind turbine	Drive train module	Gearbox	Bearings	Carrier bearing
Wind turbine	Drive train module	Gearbox	Bearings	Planet bearing
Wind turbine	Drive train module	Gearbox	Bearings	Shaft bearing
Wind turbine	Drive train module	Gearbox	Cooling system	Hose
Wind turbine	Drive train module	Gearbox	Cooling system	Pump
Wind turbine	Drive train module	Gearbox	Cooling system	Radiator
Wind turbine	Drive train module	Gearbox	Gears	Hollow shaft
Wind turbine	Drive train module	Gearbox	Gears	Planet carrier
Wind turbine	Drive train module	Gearbox	Gears	Planet gear
Wind turbine	Drive train module	Gearbox	Gears	Ring gear
Wind turbine	Drive train module	Gearbox	Gears	Spur gear
Wind turbine	Drive train module	Gearbox	Gears	Sun gear
Wind turbine	Drive train module	Gearbox	Housing	Bushing
Wind turbine	Drive train module	Gearbox	Housing	Case
Wind turbine	Drive train module	Gearbox	Housing	Mounting
Wind turbine	Drive train module	Gearbox	Housing	Torque arm system
Wind turbine	Drive train module	Gearbox	Lubrication system	Hose
Wind turbine	Drive train module	Gearbox	Lubrication system	Motor
Wind turbine	Drive train module	Gearbox	Lubrication system	Motor
Wind turbine	Drive train module	Gearbox	Lubrication system	Primary filter
Wind turbine	Drive train module	Gearbox	Lubrication system	Pump
Wind turbine	Drive train module	Gearbox	Lubrication system	Reservoir
Wind turbine	Drive train module	Gearbox	Lubrication system	Seal
Wind turbine	Drive train module	Gearbox	Lubrication system	Secondary filter
Wind turbine	Drive train module	Gearbox	Sensors	Debris
Wind turbine	Drive train module	Gearbox	Sensors	Oil level
Wind turbine	Drive train module	Gearbox	Sensors	Pressure 1

(Continues)

(Continued)

System	Sub-system	Assembly	Sub-assembly	Component
Wind turbine	Drive train module	Gearbox	Sensors	Pressure 2
Wind turbine	Drive train module	Gearbox	Sensors	Temperature
Wind turbine	Drive train module	Generator	Cooling system	Cooling fan
Wind turbine	Drive train module	Generator	Cooling system	Filter
Wind turbine	Drive train module	Generator	Cooling system	Hose
Wind turbine	Drive train module	Generator	Cooling system	Radiator
Wind turbine	Drive train module	Generator	Lubrication system	Pump
Wind turbine	Drive train module	Generator	Lubrication system	Reservoir
Wind turbine	Drive train module	Generator	Rotor	Commutator
Wind turbine	Drive train module	Generator	Rotor	Exciter
Wind turbine	Drive train module	Generator	Rotor	Resistance controller
Wind turbine	Drive train module	Generator	Rotor	Rotor lamination
Wind turbine	Drive train module	Generator	Rotor	Rotor winding
Wind turbine	Drive train module	Generator	Rotor	Slip ring
Wind turbine	Drive train module	Generator	Sensors	Core temperature sensor
Wind turbine	Drive train module	Generator	Sensors	Encoder
Wind turbine	Drive train module	Generator	Sensors	Wattmeter
Wind turbine	Drive train module	Generator	Stator	Magnet
Wind turbine	Drive train module	Generator	Stator	Stator lamination
Wind turbine	Drive train module	Generator	Stator	Stator winding
Wind turbine	Drive train module	Generator	Structural and mechanical	Front bearing
Wind turbine	Drive train module	Generator	Structural and mechanical	Housing
Wind turbine	Drive train module	Generator	Structural and mechanical	Rear bearing
Wind turbine	Drive train module	Generator	Structural and mechanical	Shaft
Wind turbine	Drive train module	Generator	Structural and mechanical	Silent block
Wind turbine	Drive train module	Main shaft set	High speed side	Coupling
Wind turbine	Drive train module	Main shaft set	High speed side	Rotor lock
Wind turbine	Drive train module	Main shaft set	High speed side	Shaft
Wind turbine	Drive train module	Main shaft set	High speed side	Transmission shaft

Wind turbine	Drive train module	Main shaft set	Low speed side	Axial bearing
Wind turbine	Drive train module	Main shaft set	Low speed side	Compression coupling
Wind turbine	Drive train module	Main shaft set	Low speed side	Connector plate
Wind turbine	Drive train module	Main shaft set	Low speed side	Main bearing seal
Wind turbine	Drive train module	Main shaft set	Low speed side	Main bearing temperature sensor
Wind turbine	Drive train module	Main shaft set	Low speed side	Main shaft
Wind turbine	Drive train module	Main shaft set	Low speed side	Radial bearing
Wind turbine	Drive train module	Main shaft set	Low speed side	Rotor lock
Wind turbine	Drive train module	Main shaft set	Low speed side	Slip ring
Wind turbine	Drive train module	Main shaft set	Mechanical brake	Calliper
Wind turbine	Drive train module	Main shaft set	Mechanical brake	Disk
Wind turbine	Drive train module	Main shaft set	Mechanical brake	Pad
Wind turbine	Drive train module	Main shaft set	Mechanical brake	Transmission lock
Wind turbine	Drive train module	Main shaft set	Sensors	High speed sensor
Wind turbine	Drive train module	Main shaft set	Sensors	Low speed sensor
Wind turbine	Drive train module	Main shaft set	Sensors	Position sensor
Wind turbine	Electrical module	Auxiliary electrical system	Electrical services	24 DC feeder
Wind turbine	Electrical module	Auxiliary electrical system	Electrical services	Auxiliary transformer
Wind turbine	Electrical module	Auxiliary electrical system	Electrical services	Breaker
Wind turbine	Electrical module	Auxiliary electrical system	Electrical services	Cabinet
Wind turbine	Electrical module	Auxiliary electrical system	Electrical services	Fan
Wind turbine	Electrical module	Auxiliary electrical system	Electrical services	Fuse
Wind turbine	Electrical module	Auxiliary electrical system	Electrical services	Grid protection relay
Wind turbine	Electrical module	Auxiliary electrical system	Electrical services	Light
Wind turbine	Electrical module	Auxiliary electrical system	Electrical services	Mechanical switch
Wind turbine	Electrical module	Auxiliary electrical system	Electrical services	Power point
Wind turbine	Electrical module	Auxiliary electrical system	Electrical services	Protection cabinet
Wind turbine	Electrical module	Auxiliary electrical system	Electrical services	Pushbutton
Wind turbine	Electrical module	Auxiliary electrical system	Electrical services	Relay
Wind turbine	Electrical module	Auxiliary electrical system	Electrical services	Space heater

(Continues)

(*Continued*)

System	Sub-system	Assembly	Sub-assembly	Component
Wind turbine	Electrical module	Auxiliary electrical system	Electrical services	Surge arrester
Wind turbine	Electrical module	Auxiliary electrical system	Electrical services	Thermal protection
Wind turbine	Electrical module	Auxiliary electrical system	Electrical services	Ups
Wind turbine	Electrical module	Auxiliary electrical system	Lightning protection system	Air termination
Wind turbine	Electrical module	Auxiliary electrical system	Lightning protection system	Bonding element
Wind turbine	Electrical module	Auxiliary electrical system	Lightning protection system	Earth connector
Wind turbine	Electrical module	Auxiliary electrical system	Lightning protection system	Earth termination
Wind turbine	Electrical module	Auxiliary electrical system	Lightning protection system	Sliding contact
Wind turbine	Electrical module	Auxiliary electrical system	Lightning protection system	Spark gap system
Wind turbine	Electrical module	Auxiliary electrical system	Lightning protection system	Surge arrester
Wind turbine	Electrical module	Control and communication system	Ancillary equipment	Breaker
Wind turbine	Electrical module	Control and communication system	Ancillary equipment	Cabinet temperature sensor
Wind turbine	Electrical module	Control and communication system	Ancillary equipment	Cable
Wind turbine	Electrical module	Control and communication system	Ancillary equipment	Contactor
Wind turbine	Electrical module	Control and communication system	Communication system	Analog I/O unit
Wind turbine	Electrical module	Control and communication system	Communication system	Digital I/O unit
Wind turbine	Electrical module	Control and communication system	Communication system	Ethernet module
Wind turbine	Electrical module	Control and communication system	Communication system	Field bus master
Wind turbine	Electrical module	Control and communication system	Communication system	Field bus slave
Wind turbine	Electrical module	Control and communication system	Communication system	Frequency unit
Wind turbine	Electrical module	Control and communication system	Condition monitoring system	Condition cables
Wind turbine	Electrical module	Control and communication system	Condition monitoring system	Data logger
Wind turbine	Electrical module	Control and communication system	Condition monitoring system	Protocol adapter card for data logger
Wind turbine	Electrical module	Control and communication system	Condition monitoring system	Sensors
Wind turbine	Electrical module	Control and communication system	Controller hardware	Controller power supply
Wind turbine	Electrical module	Control and communication system	Controller hardware	CPU
Wind turbine	Electrical module	Control and communication system	Controller hardware	Internal communication system

Wind turbine	Electrical module	Control and communication system	Controller hardware	Main I/O unit
Wind turbine	Electrical module	Control and communication system	Controller hardware	Watch dog unit
Wind turbine	Electrical module	Control and communication system	Controller software	Closed loop control software
Wind turbine	Electrical module	Control and communication system	Controller software	Supervisory control software
Wind turbine	Electrical module	Control and communication system	Safety chain	Emergency button
Wind turbine	Electrical module	Control and communication system	Safety chain	Max speed switch
Wind turbine	Electrical module	Control and communication system	Safety chain	Power switch
Wind turbine	Electrical module	Control and communication system	Safety chain	Short circuit switch
Wind turbine	Electrical module	Control and communication system	Safety chain	Vibration switch
Wind turbine	Electrical module	Control and communication system	Safety chain	Watch dog switch
Wind turbine	Electrical module	Control and communication system	Safety chain	Wind-up switch
Wind turbine	Electrical module	Frequency converter	Converter auxiliaries	Auxiliary power supply
Wind turbine	Electrical module	Frequency converter	Converter auxiliaries	Cabinet
Wind turbine	Electrical module	Frequency converter	Converter auxiliaries	Cabinet heating system
Wind turbine	Electrical module	Frequency converter	Converter auxiliaries	Cabinet sensor
Wind turbine	Electrical module	Frequency converter	Converter auxiliaries	Communication and interface unit
Wind turbine	Electrical module	Frequency converter	Converter auxiliaries	Control board
Wind turbine	Electrical module	Frequency converter	Converter auxiliaries	Generator side fan
Wind turbine	Electrical module	Frequency converter	Converter auxiliaries	Grid side fan
Wind turbine	Electrical module	Frequency converter	Converter auxiliaries	Measurement unit
Wind turbine	Electrical module	Frequency converter	Converter auxiliaries	Power supply
Wind turbine	Electrical module	Frequency converter	Converter auxiliaries	Power supply 24 V
Wind turbine	Electrical module	Frequency converter	Converter auxiliaries	Tachometer adapter
Wind turbine	Electrical module	Frequency converter	Converter auxiliaries	Thermostat
Wind turbine	Electrical module	Frequency converter	Converter power bus	Branching unit
Wind turbine	Electrical module	Frequency converter	Converter power bus	Capacitor
Wind turbine	Electrical module	Frequency converter	Converter power bus	Contactor
Wind turbine	Electrical module	Frequency converter	Converter power bus	Generator side converter
Wind turbine	Electrical module	Frequency converter	Converter power bus	Generator side power module
Wind turbine	Electrical module	Frequency converter	Converter power bus	Grid side converter

(*Continues*)

(*Continued*)

System	Sub-system	Assembly	Sub-assembly	Component
Wind turbine	Electrical module	Frequency converter	Converter power bus	Grid side power module
Wind turbine	Electrical module	Frequency converter	Converter power bus	Inductor
Wind turbine	Electrical module	Frequency converter	Converter power bus	Load switch
Wind turbine	Electrical module	Frequency converter	Converter power bus	Pre-charge unit
Wind turbine	Electrical module	Frequency converter	Power conditioning	Common mode filter
Wind turbine	Electrical module	Frequency converter	Power conditioning	Crowbar
Wind turbine	Electrical module	Frequency converter	Power conditioning	DC chopper
Wind turbine	Electrical module	Frequency converter	Power conditioning	Generator side filter
Wind turbine	Electrical module	Frequency converter	Power conditioning	Line filter assembly
Wind turbine	Electrical module	Frequency converter	Power conditioning	Voltage limiter unit
Wind turbine	Electrical module	Power electrical system	Measurements	
Wind turbine	Electrical module	Power electrical system	Measurements	
Wind turbine	Electrical module	Power electrical system	Power circuit	Cables
Wind turbine	Electrical module	Power electrical system	Power circuit	Machine contactor
Wind turbine	Electrical module	Power electrical system	Power circuit	Machine transformer
Wind turbine	Electrical module	Power electrical system	Power circuit	MV busbar/isolator
Wind turbine	Electrical module	Power electrical system	Power circuit	MV switchgear
Wind turbine	Electrical module	Power electrical system	Power circuit	Soft start electronics
Wind turbine	Nacelle module	Hydraulics system	Hydraulic power pack	Motor
Wind turbine	Nacelle module	Hydraulics system	Hydraulic power pack	Pump
Wind turbine	Nacelle module	Hydraulics system	Hydraulic power pack	Pressure valve
Wind turbine	Nacelle module	Hydraulics system	Hydraulic power pack	Filter
Wind turbine	Nacelle module	Hydraulics system	Actuator	Bushing
Wind turbine	Nacelle module	Hydraulics system	Actuator	Cylinder
Wind turbine	Nacelle module	Hydraulics system	Actuator	Hose/Fitting
Wind turbine	Nacelle module	Hydraulics system	Actuator	Hydraulic linear drive
Wind turbine	Nacelle module	Hydraulics system	Actuator	Limit switch
Wind turbine	Nacelle module	Hydraulics system	Actuator	Linkage

Wind turbine	Nacelle module	Hydraulics system	Actuator	Miscellaneous hydraulics system
Wind turbine	Nacelle module	Hydraulics system	Actuator	Position controller
Wind turbine	Nacelle module	Hydraulics system	Actuator	Proportional valve
Wind turbine	Nacelle module	Hydraulics system	Actuator	Pump
Wind turbine	Nacelle module	Hydraulics system	Torque converter	
Wind turbine	Nacelle module	Hydraulics system	Differential	
Wind turbine	Nacelle module	Hydraulics system	Viscous coupling	
Wind turbine	Nacelle module	Nacelle auxiliaries	Meteorological sensors	Anemometer
Wind turbine	Nacelle module	Nacelle auxiliaries	Meteorological sensors	Wind vane
Wind turbine	Nacelle module	Nacelle auxiliaries	Nacelle sensors	Emergency vibration sensor
Wind turbine	Nacelle module	Nacelle auxiliaries	Nacelle sensors	Yaw encoder
Wind turbine	Nacelle module	Nacelle auxiliaries	Safety system	Beacon
Wind turbine	Nacelle module	Nacelle auxiliaries	Safety system	Down conductor
Wind turbine	Nacelle module	Nacelle auxiliaries	Safety system	Fall arrester
Wind turbine	Nacelle module	Nacelle auxiliaries	Safety system	Fire fighting system
Wind turbine	Nacelle module	Nacelle auxiliaries	Safety system	Nacelle cover metallic mesh
Wind turbine	Nacelle module	Nacelle auxiliaries	Safety system	Lightning protection termination
Wind turbine	Nacelle module	Nacelle auxiliaries	Safety system	Service crane
Wind turbine	Nacelle module	Nacelle structure	Bedplate	Bolts
Wind turbine	Nacelle module	Nacelle structure	Bedplate	Cast or welded structure
Wind turbine	Nacelle module	Nacelle structure	Cover	Fibreglass
Wind turbine	Nacelle module	Nacelle structure	Cover	Hatch
Wind turbine	Nacelle module	Nacelle structure	Generator frame	Bolts
Wind turbine	Nacelle module	Nacelle structure	Generator frame	Cast or welded structure
Wind turbine	Nacelle module	Yaw system	Yaw brake	Yaw brake callipers
Wind turbine	Nacelle module	Yaw system	Yaw brake	Yaw brake disc
Wind turbine	Nacelle module	Yaw system	Yaw brake	Yaw brake hoses
Wind turbine	Nacelle module	Yaw system	Yaw brake	Yaw brake paths
Wind turbine	Nacelle module	Yaw system	Yaw drive	Damper

(*Continues*)

(*Continued*)

System	Sub-system	Assembly	Sub-assembly	Component
Wind turbine	Nacelle module	Yaw system	Yaw drive	Yaw bearing
Wind turbine	Nacelle module	Yaw system	Yaw drive	Yaw gearbox
Wind turbine	Nacelle module	Yaw system	Yaw drive	Yaw motor
Wind turbine	Nacelle module	Yaw system	Yaw drive	Yaw pinion
Wind turbine	Nacelle module	Yaw system	Yaw sensors	Wind-up counter
Wind turbine	Nacelle module	Yaw system	Yaw sensors	Yaw encoder
Wind turbine	Rotor module	Blade	Blade lightning protection termination	Blade lightning protection termination
Wind turbine	Rotor module	Blade	Blade lightning down-conductor	Blade lightning down-conductor
Wind turbine	Rotor module	Blade	De-icing system	De-icing system
Wind turbine	Rotor module	Blade	Leading edge bond	Leading edge bond
Wind turbine	Rotor module	Blade	Nuts and bolts	Nuts and bolts
Wind turbine	Rotor module	Blade	Paint and coating	Paint and coating
Wind turbine	Rotor module	Blade	Root structure	Root structure
Wind turbine	Rotor module	Blade	Sandwich shell	Sandwich shell
Wind turbine	Rotor module	Blade	Spar box	Spar box
Wind turbine	Rotor module	Blade	Spar cap	Spar cap
Wind turbine	Rotor module	Blade	Spar web	Spar web
Wind turbine	Rotor module	Blade	Trailing edge bond	Trailing edge bond
Wind turbine	Rotor module	Hub	Exit hatch	Exit hatch
Wind turbine	Rotor module	Hub	Nose cone	Nose cone
Wind turbine	Rotor module	Pitch system	Pitch cabinet	Battery
Wind turbine	Rotor module	Pitch system	Pitch cabinet	Battery charger
Wind turbine	Rotor module	Pitch system	Pitch cabinet	Heater
Wind turbine	Rotor module	Pitch system	Pitch cabinet	Local controller
Wind turbine	Rotor module	Pitch system	Pitch cabinet	Switchboard
Wind turbine	Rotor module	Pitch system	Pitch drive	Motor

Wind turbine	Rotor module	Pitch system	Pitch drive	Motor cooling
Wind turbine	Rotor module	Pitch system	Pitch drive	Motor cooling system
Wind turbine	Rotor module	Pitch system	Pitch drive	Motor drive
Wind turbine	Rotor module	Pitch system	Pitch drive	Pinion
Wind turbine	Rotor module	Pitch system	Pitch drive	Pitch bearing
Wind turbine	Rotor module	Pitch system	Pitch drive	Pitch gearbox
Wind turbine	Rotor module	Pitch system	Pitch sensors	Position encoder
Wind turbine	Rotor module	Pitch system	Pitch sensors	Temperature sensor
Wind turbine	Rotor module	Pitch system	Pitch sensors	Voltmeter
Wind turbine	Support structure	Foundation	Gravity-based foundation	Concrete
Wind turbine	Support structure	Foundation	Gravity-based foundation	Steel reinforcement
Wind turbine	Support structure	Foundation	Monopile	Corrosion protection
Wind turbine	Support structure	Foundation	Monopile	Pile
Wind turbine	Support structure	Foundation	Monopile	Transition piece
Wind turbine	Support structure	Foundation	Onshore	Concrete
Wind turbine	Support structure	Foundation	Onshore	Nuts and bolts
Wind turbine	Support structure	Foundation	Onshore	Piles
Wind turbine	Support structure	Foundation	Onshore	Steel reinforcement
Wind turbine	Support structure	Foundation	Space frame/tripod	Corrosion protection
Wind turbine	Support structure	Foundation	Space frame/tripod	Piles
Wind turbine	Support structure	Foundation	Space frame/tripod	Structures
Wind turbine	Support structure	Tower	Access equipment	Ladder
Wind turbine	Support structure	Tower	Access equipment	Landing pad
Wind turbine	Support structure	Tower	Access equipment	Lightning protection
Wind turbine	Support structure	Tower	Tower	Climb assist
Wind turbine	Support structure	Tower	Tower	Maintenance crane
Wind turbine	Support structure	Tower	Tower	Nuts and bolts
Wind turbine	Support structure	Tower	Tower	Paint/coating
Wind turbine	Support structure	Tower	Tower	Tower section
Wind farm	Collection system	Cable	Cable	Cable

(Continues)

(Continued)

System	Sub-system	Assembly	Sub-assembly	Component
Wind farm	Meteorological station	Meteorological station	Meteorological station	Meteorological station
Wind farm	Operational infrastructure	Operational infrastructure	Operational infrastructure	Operational infrastructure
Wind farm	Substation	Grid connection	HV link	HV link
Wind farm	Substation	Grid connection	Substation transformer	Substation transformer
Wind farm	Substation	Grid connection	Utility communication and control	Utility communication and control

11.7   Detailed wind turbine failure terminology

Sub-system	Assembly	Sub-assembly or component	Failure or failure mode from ReliaWind WP2
Foundation	Monopile		Scour; erosion; corrosion
	Tripod		Erosion; corrosion
	Gravity base		Scour; erosion
	Transition piece		Grout slippage; grout loss
	Jacking brackets		Fatigue
	Bolts		Fatigue; corrosion; erosion
Tower	Structure		Fatigue; corrosion
	Bolts		Corrosion; overload; fatigue
	Climbing system		Corrosion; overload
	Lift		Motor failure; interlock failure
Rotor module	Rotor	Rotor hub	Fracture; corrosion
		Rotor blade	Mechanical imbalance; aerodynamic imbalance
		Spars	Cracking; debonding from skin
		Coating	Roughening; impact damage
		Lamination	Debonding; lightning damage; impact damage
		Leading edge	Erosion; ice build-up
		Trailing edge	Debonding; ice build-up
		Tip brake	Loss of tip
		Tip brake wire	Snagging; broken
Rotor module	Pitch system		Pitch bearing failure; seizure; overload; motor failure
	Pitch system		Hydraulic oil contamination; hydraulic oil leakage; hydraulic pump failure
	Pitch system		Slip ring wear
	Pitch system		Blade mismatch; aerodynamic imbalance
Nacelle module	Yaw system		Yaw bearing failure; yaw ring wear; yaw ring distortion or damage; yaw motor failure; yaw brake failure; yaw brake seizure; yaw alignment error
	Hydraulics system	Hydraulic power pack; motor	Winding failure; over-temperature
		Hydraulic power pack; pump	Over-temperature; seal failure
		Hydraulic power pack; pressure valve	Seal failure
		Hydraulic power pack; filter	Blockage
	Anemometer		Ice build-up; seizure; calibration drift; impact damage

(Continues)

(*Continued*)

Sub-system	Assembly	Sub-assembly or component	Failure or failure mode from ReliaWind WP2
	Wind vane		Ice build-up; seizure; calibration drift; impact damage
	Electrical system		See sub-systems
	Access system		Wear; looseness; breakage
	Generator supports		Cracking; bending; looseness
Drive train	Main bearing		Bearing failure; misalignment; lubrication
	Main shaft		Cracking; permanent bend
	Mechanical brake		Pad wear; overheating; disk wear; hydraulic failure
	Gearbox	Gear case	Fracture
		Suspension	Wear; looseness
		Torque arm	Wear; looseness
		Lubrication system	Loss of lubricant; contaminated lubricant; aged lubricant; lubricant system failure; lubrication pump failure; blocked lubrication filters; blocked jets
		Epicyclical part, planet carrier	Lubrication
		Epicyclical part, planet bearing	Bearing failure; lubrication
		Epicyclical part, planet gear	Tooth failure; lubrication
		Epicyclical part, internal gear	Tooth failure; lubrication; fracture
		Epicyclical part, sun gear	Tooth failure; lubrication
		Epicyclical part, shaft	Cracking; journal damage
		Parallel shaft part, gear	Tooth failure; lubrication
		Parallel shaft part, bearing	Bearing failure; lubrication
		Parallel shaft part, pinion	Tooth failure; lubrication
		Parallel shaft part, shaft	Cracking; journal damage
		High-speed shaft	Cracking; permanent bend
	Generator	Coupling	Misalignment; perishing; wear
		Rotor	Fracture
		Rotor windings	Shorted turn; earth failure; broken bar
		Stator	Shorted turn; earth failure
		Stator windings	Shorted turn; earth failure
		Bearings	Bearing failure; lubrication failure Blockage; over-temperature

(*Continued*)

Sub-system	Assembly	Sub-assembly or component	Failure or failure mode from ReliaWind WP2
Electrical module	Frequency converter	Stator cooling system	
		Slip rings	Brush wear; over-temperature
		Encoder	Encoder failure
		Power electronics	Component failure; joint failure
		Grid-side filter	Component failure
		Grid-side inverter	IGBT failure; over-temperature
		DC link	Capacitor failure
		Generator-side inverter	IGBT failure; over-temperature
		Generator-side filter	Component failure
		Crowbar	Thyristor failure
		Crowbar resistor	Over-temperature; component failure; fuse failure
		DC chopper	Component failure
		DC chopper resistor	Over-temperature; component failure; fuse failure
	Transformer	Windings	Winding failure; over-temperature
		Core	Over-temperature
		Oil system	Oil deterioration; over-temperature
	Switchgear		Circuit breaker failure

11.8   References

[1]   IEC 61400-3:2010 draft. Wind turbines – Design requirements for offshore wind turbines. International Electrotechnical Commission

[2]   OREDA, offshore reliability data:
- OREDA-1984. *Offshore Reliability Data Handbook*. 1st edn. VERITEC – Marine Technology Consultants, PennWell Books
- OREDA-1997. *Offshore Reliability Data Handbook*. 3rd edn. SINTEF Industrial Management. Det Norske Veritas, Norway
- OREDA-2002. *Offshore Reliability Data Handbook*. 4th edn. SINTEF Industrial Management. Det Norske Veritas, Norway

[3]   EN ISO 14224:2006, Petroleum, petrochemical and natural gas industries – Collection and exchange of reliability and maintenance data for equipment

[4]   Faulstich S., Durstewitz M., Hahn B., Knorr K., Rohrig K. Windenergie Report. Institut für solare Energieversorgungstechnik, Kassel, Deutschland; 2008

[5]   Schmid J., Klein H.P. *Performance of European Wind Turbines*. London and New York: Elsevier, Applied Science; 1991. ISBN 1-85166-737-7

[6]   Arabian-Hoseynabadi H., Oraee H., Tavner P.J. 'Failure modes and effects analysis (FMEA) for wind turbines'. *International Journal of Electrical Power & Energy Systems*. 2010;**32**(7):817–24

[7]   IEC 61400-25-6:2010. Wind turbines – Communications for monitoring and control of wind power plants – Logical node classes and data classes for condition monitoring. International Electrotechnical Commission

[8]   VGB PowerTech, Guideline, Reference Designation System for Power Plants, RDS-PP, Application Explanation for Wind Power Plants, VGB-B 116 D2. 1st edn. 2007

Chapter 12

Appendix 3: WMEP operators report form

| ******* (Company name) Maintenance and repair report | Work done on: Date | Month | Year | Report No: |

Post code / Plant ID number

Cause work

- Scheduled maintenance (only examination and functional check)
- Scheduled maintenance with replacement of worn components or repair of faults
- Unscheduled maintenance or repair after malfunctions

Down times

- Not stopped
- Stopped

From
To
Date   Month   Year

Reading of hour counter

Cost according to calculation

- Material £
- Labour £
- Journey £
- Total cost (incl. tax) £

Comments

The Operator

Place/Date

Signature

Cause of malfunction

- High wind
- Grid failure
- Lightning
- Icing
- Malfunction of control system
- Component wear or failure
- Component loosening
- Other cause
- Unknown cause

Effect of malfunction

- Overspeed
- Overload
- Noise
- Vibrations
- Reduced power output
- Causing follow up damages
- Plant stoppage
- Other consequences

Removal of malfunction

Faultless operation without later repair:

- Control reset
- Changing control parameters

Repaired or replaced components:

- **Rotor hub**
  - Hub body
  - Pitch mechanism
  - Pitch bearing
- **Rotor blades**
  - Blade bolts
  - Blade shell
  - Aerodynamic brakes
- **Generator**
  - Windings
  - Brushes
  - Bearings
- **Electrical system**
  - Inverter
  - Fuses
  - Switches
  - Cables/connections
- **Sensors**
  - Anemometer/Wind vane
  - Vibration switch
  - Temperature switch
  - Oil pressure switch
  - Power sensor
  - Rev counter
- **Control system**
  - Electronic control unit
  - Relay
  - Measurement cables and connections

- **Gear box**
  - Bearings
  - Gear-wheels
  - Gear shaft
  - Sealings
- **Mechanical brakes**
  - Brake disc
  - Brake pads
  - Brake shoe
- **Drive train**
  - Rotor bearings
  - Drive shafts
  - Couplings
- **Hydraulic system**
  - Hydraulic pump
  - Pump motor
  - Valves
  - Hydraulic pipes/hoses
- **Yaw system**
  - Yaw bearings
  - Yaw motor
  - Wheels and pinions
- **Structural components/Housing**
  - Foundation
  - Tower/Tower bolts
  - Nacelle frame
  - Nacelle cover
  - Ladder/lift

Main component exchanged

Please check if complete component is exchanged

- Nacelle
- Rotor blades
- Rotor hub
- Gear box
- Generator
- Yaw system
- Tower
- Control system cabinet
- Grid transformer

[Source: Reference 4 of Chapter 11]

Chapter 13

Appendix 4: Commercially available SCADA systems for WTs

13.1 Introduction

A wind farm's existing SCADA data stream is a valuable resource, which can be monitored by WT OEMs, operators and other experts to observe, and hence optimise the performance of the WT. In order to conduct an efficient SCADA data analysis, data analysis tools are required.

This survey discusses commercially available SCADA systems that are currently being applied in the WT Industry.

13.2 SCADA data

SCADA systems are a standard installation in large WTs and wind farms – their data being collected from individual WT controllers. According to Reference 1, the SCADA system assesses the status of the WT and its sub-systems using sensors fitted to the WT, such as anemometers, thermocouples and switches. The signals from these instruments are monitored and recorded at a low data rate, usually at 10 minute intervals. The SCADA data shows the operating condition of a WT. Many large WTs are now fitted with CMSs, which monitor sensors associated with the rotating drive train, such as accelerometers, proximeters and oil particle counters. The CMS is normally separate from the SCADA and collects data at much higher data rates.

By analysing SCADA data, we are able to observe the relationship between different signals, and hence deduce the health of WT sub-assemblies. It would prove beneficial, from the perspective of utility companies, if the data could be analysed and interpreted automatically to support the operators in identifying defects.

13.3 Commercially available SCADA data analysis tools

Table 13.1 provides a summary of the available SCADA systems based on information collected from Internet. It should be noted that the table is accurate up to 2011

Table 13.1   Commercially available SCADA systems

Ref	Product and company information			Product details	
	Product name	**Company**	**Country of origin**	**Description**	**Main function**
1	BaxEnergy WindPower Dashboard [2]	BaxEnergy GmbH	Germany	BaxEnergy WindPower Dashboard offers an extensive and comprehensive customisation for full integration of SCADA system applications and SCADA software.	Real-time data acquisition and visualization, alarm analytics and data reporting
2	CitectSCADA[3]	Schneider Electric Pty. Ltd.	Australia	CitectSCADA is a reliable, flexible and high performance system for any industrial automation monitoring and control application.	Graphical process visualisation, superior alarm management, built-in reporting and powerful analysis tools
3	CONCERTO [4]	AVL	Austria	CONCERTO is a commercially available analysis and post-processing system capable of handling large quantities of data.	One tool for manifold applications, all data post-processing tasks within one tool, advanced data management
4	ENERCON SCADA system [5]	ENERCON	USA	The Enercon SCADA system is used for data acquisition, remote monitoring, open-loop and closed-loop control for both individual WT and wind farms. It enables the customer and Enercon service to monitor the operating state and to analyse saved operating data.	Requesting status data, storing operating data, wind farm communication and loop control of the wind farm
5	Gamesa WindNet [6]	Gamesa	Spain	The WindNet SCADA system in a wind farm is configured with a basic hardware and software platform based on Windows technology. The user interface is an easy to use SCADA application with specific options for optimal supervision and control of a wind farm including devices like WTs, meteorological masts and a substation.	Supervision and control of WTs and meteorological masts, alarm and warning management, report generation and user management

6	GE – HMI/SCADA – iFIX 5.1 [7]	GE (General Electric Co.)	USA	iFIX is a superior proven real-time information management and SCADA solution. It is open, flexible and scalable, which includes impressive next-generation visualisations, a reliable control engine, a powerful built-in historian and more.	Real-time data management and control, information analysis
7	GH SCADA [8]	GL Garrad Hassan	Germany	GH SCADA has been designed by Garrad Hassan in collaboration with WT OEMs, wind farm operators, developers and financiers to meet the needs of all those involved in wind farm operation, analysis and reporting.	Remote control of individual WT, online data viewing, reports and analysis
8	ICONICS for Renewable Energy [8]	ICONICS Inc.	USA	ICONICS provides portals for complete operations, including energy analytics, data histories and reports, GEO SCADA with meteorological updates.	Portals for complete operations, data histories and reports, GEO SCADA with meteorological updates
9	InduSoft Wind Power solutions [9]	InduSoft	USA	InduSoft Web Studio software brings you a powerful HMI/SCADA package that can monitor and adjust any operating set point in the controller or PLC.	WT monitoring, maintenance assistance and control room
10	INGESYS Wind IT [10]	IngeTeam	Spain	INGESYS Wind IT makes it possible to completely integrate all the wind power plants into a single system. It provides advanced reporting services.	Advanced reporting, client/server architecture, standard protocols and formats
11	reSCADA [11]	Kinetic Automation Pty. Ltd.	USA	reSCADA targets and specialises in renewable energy industries. It saves time, effort and cost in developing HMI/SCADA.	Office 2007 GUI style, data visualization, diary and mapping tools

(Continues)

Table 13.1 (*Continued*)

Ref	Product and company information			Product details	
	Product name	Company	Country of origin	Description	Main function
12	SgurrTREND [13]	SgurrEnergy	UK	SgurrEnergy provides a variety of wind monitoring solutions to evaluate the wind resource potential at your prospective wind farm site, offering a one-stop shop for all mast services from planning application, data collection and mast decommissioning to wind analysis services for energy yield prediction, project layout and design services.	Wind monitoring, processing and archiving the data and reporting services
13	SIMAP (Reference 10 of Chapter 7)	Molinos del Ebro, S.A.	Spain	SIMAP is based on artificial intelligence techniques. It is able to create and dynamically adapt a maintenance calendar for the WT that is monitoring. The new and positive aspects of this predictive maintenance methodology have been tested in WTs.	Continuous collection of data, continuous processing information, failure risk forecasting and dynamical maintenance scheduling
14	WindCapture [14]	SCADA solutions	Canada	WindCapture is a SCADA software package used for monitoring, controlling and data collection and reporting for WT generators. It was designed and tailored to the demands of OEMs, operators, developers and maintenance managers of wind energy project and facilities.	Real-time data reporting with the highest degree of accuracy, advanced GUI

| 15 | Wind Systems [15] | SmartSignal | USA | SmartSignal analyses real-time data and detects and notifies wind farms of impending problems, allowing owners to focus on fixing problems early and efficiently. | Model maintenance, monitoring services and predictive diagnostics |
| 16 | Wind Asset Monitoring Solution [12] | Matrikon | Canada | The solution bridges the gap between instrumentation and management systems to enable and sustain operational excellence by retrieving and better managing data that is not often readily accessible. | Monitor and manage all remote assets, leverage and integrate with SCADA and CMMS |

but may not be definitive. The products are arranged alphabetically by product name. A quick summary of Table 13.1 shows that:

- three products are developed in association with WT OEMs (4, 5 and 6);
- two products are developed by renewable energy consultancies (7 and 12);
- nine products are developed by industrial software companies including manufacturers of the WT controllers (1, 2, 3, 8, 9, 11, 14, 15 and 16);
- one product is developed by WT operator (13);
- one product is developed by an electrical equipment provider (10).

Among these 16 products, Gamesa WindNet (5) and Enercon SCADA system (4) are wind farm cluster management systems. Both provide a framework for data acquisition, remote monitoring, open/closed loop control and data analysis for both individual WTs and wind farms. The Enercon SCADA system was launched in 1998 and is now used in conjunction with more than 11,000 WTs. Gamesa WindNet consists of a wide area network (WAN) system for wind farms connected to an operational centre.

GE – HMI/SCADA – iFIX 5.1 (6) was developed by General Electric Co. (GE), also a WT OEM. It is ideally suited for complex SCADA applications. The software also enables faster, better intelligent control and visibility of wind farm operations.

GH SCADA (7) and SgurrTREND (12) were developed by renewable energy consultancies in collaboration with WT OEMs, wind farm operators, developers and financiers to meet the needs of all those involved in wind farm operation, analysis and reporting.

CONCERTO (3) is not specialised for SCADA data analysis. It is a generic data post-processing tool focusing on quick and intuitive signal analysis, validation, correlation and reporting for any kind of acquired data. Gray and Watson used it to perform analysis of WT SCADA data (Reference 9 of Chapter 7).

SIMAP (13) is based on artificial intelligence techniques. The new and positive aspects of this predictive maintenance methodology have been tested on WTs. SIMAP has been applied to a wind farm owned by a Spanish wind energy company called Molinos del Ebro, S.A. (Reference 10 of Chapter 7).

INGESYS Wind IT (10) was developed by IngeTeam, an electrical equipment provider. The system aims to integrate wind power plants into a single system and then optimise wind farm management. INGENSYS Wind IT also provides an advanced reporting service for power curve analysis, faults, alarms and customer reports.

The other products – BaxEnergy WindPower Dashboard (1), CitectSCADA (2), ICONICS for Renewable Energy (8), InduSoft Wind Power (9), reSCADA (11), WindCapture (14), Wind Systems (15), MATRIKON Wind Asset Monitoring Solution (16) – were developed by industrial software companies, which integrated SCADA systems to provide a reliable, flexible and high performance application for WT automation, monitoring and control.

13.4 Summary

From this survey we can conclude that:

- There is a wide variety of commercial SCADA systems available to the wind industry.
- Most of the commercially available SCADA systems are able to analyse real-time data.
- The performance analysis techniques used in SCADA systems vary from tailored statistical method to artificial intelligence.
- Successful SCADA systems provide cluster management for wind farms. They provide a framework for data acquisition, alarm management, reporting and analysis, production forecasting and meteorological updates.
- Some built-in diagnostics techniques are able to diagnose the sub-assembly failure of WT.
- Finally, it should be noted that the development of SCADA systems is aimed to provide a reliable, flexible and high performance system for WT automation monitoring and control. The industry is already noting the importance of operational parameters such as load and speed and so techniques may begin to adapt further to the WT environment leading to more reliable WT diagnostics solution.

13.5 References

[1] Wind on the Grid. Available from http://www.windgrid.eu/Deliverables_EC/D6%20WCMS.pdf

[2] BAX Energy. Available from http://www.baxenergy.com/integration.htm [Accessed 18 October 2010]

[3] CitectSCADA. Available from http://www.citect.com/index.php?option=com_content&view=article&id=1457&Itemid=1314 [Accessed 23 September 2010]

[4] CONCERTO. Available from https://www.avl.com/concerto [Accessed 15 October 2010]

[5] ENERCON SCADA system. Available from http://www.enercon-eng.com/ or http://www.windgrid.eu/Deliverables_EC/D6%20WCMS.pdf [Accessed 23 September 2010]

[6] Gamesa WindNet. Available from http://www.gamesacorp.com/en/products/wind-turbines/design-and-development/gamesa-windnet or http://www.windgrid.eu/Deliverables_EC/D6%20WCMS.pdf [Accessed on 23 September 2010]

[7] GE – HMI/SCADA. Available from http://www.ge-ip.com/products/3311?cid=GlobalSpec [Accessed 23 September 2010]

[8] GH SCADA. Available from http://www.gl-garradhassan.com/en/software/scada.php [Accessed 23 September 2010]; ICONICS for Renewable Energy.

Available from http://www.iconics.com/industries/renewable.asp [Accessed 23 September 2010]

 [9]   InduSoft Wind Power solutions. Available from http://www.indusoft.com/PDF/wind_brochure_090803b.pdf [Accessed 23 September 2010]

[10]   INGESYS Wind IT from IngeTeam Integrated Wind Farm Management System. Available from http://pdf.directindustry.com/pdf/ingeteam-power-technology-sa/ingesys-it-scada-technical-information/62841-117937.html [Accessed 25 May 2012]

[11]   reSCADA. Available from http://www.kineticautomation.com/pc.html [Accessed 23 September 2010]

[12]   MATRIKON™ Wind Asset Monitoring Solution. Available from http://www.matrikon.com/power/wind.aspx [Accessed 29 October 2010]

[13]   SgurrTREND. Available from http://www.sgurrenergy.com/Services/Full LifeCycle/windMonitoring.php [Accessed 23 September 2010]

[14]   WindCapture. Available from http://www.scadasolutions.com/livesite/prods-scada.shtml [Accessed 23 September 2010]

[15]   Wind Systems. Available from http://www.smartsignal.com/industries/wind.aspx [Accessed 23 September 2010]

Chapter 14

Appendix 5: Commercially available condition monitoring systems for WTs

14.1 Introduction

As wind energy assumes greater importance in remote and offshore locations, effective and reliable condition monitoring (CM) techniques are required. Conventional CM methods used in the power generation industry have been adapted by a number of industrial companies and have been applied to WTs commercially.

This survey considers commercially available condition monitoring systems (CMSs) currently applied in the wind industry. Information has been gathered over several years from conferences and websites and includes information available from product brochures, technical documents and discussion with company representatives. The research was carried out as part of the Supergen Wind Energy Technologies Consortium whose objective is to devise a comprehensive CMS for practical application on WTs. The survey also indentifies some of the advantages and disadvantages of existing commercial CMSs alongside discussion of access, cost, connectivity and commercial issues surrounding the application of WT CMSs.

14.2 Reliability of wind turbines

Quantitative studies of WT reliability have recently been carried out based on publicly available data referred to in Chapter 3. These studies have shown WT gearboxes to be a mature technology with constant or slightly deteriorating reliability with time. This would suggest that WT gearboxes are not an issue; however, surveys by WMEP and LWK [1, 2] have shown that gearboxes exhibit the highest downtime per failure among onshore sub-assemblies. This is shown graphically in Figure 3.5 where we clearly see a consistently low gearbox failure rate between two surveys with high downtime per failure. Similar results have also been shown for the Egmond aan Zee wind farm [3] where gearbox failure rate is not high but the downtime and resulting costs are. The poor early reliabilities for gearbox and drive train reliability components have led to an emphasis in WT CMSs on drive train components and therefore on vibration analysis.

The high downtime for gearboxes derives from complex repair procedures. Offshore WT maintenance can be a particular problem as this involves specialist equipment such as support vessels and cranes but has the additional issue of potentially unfavourable weather and wave conditions. The EU-funded project ReliaWind [4] developed a systematic and consistent process to deal with detailed commercial data collected from operational wind farms. This includes the analysis of 10 minute average SCADA data as discussed earlier, automated fault logs and operation and maintenance reports. However, more recent information on WT reliability and downtime, especially when considering offshore operations, suggests that the target for WT CMSs should be widened from the drive train towards WT electrical and control systems [5].

As a result of low early WT reliability, particularly in larger WTs and as a result of the move offshore, interest in CMS has increased. This is being driven forward by the insurer Germanischer Lloyd who published guidelines for the certification of CMSs [6] and certification of WTs both onshore [7] and offshore [8].

14.3   Monitoring of wind turbines

WTs are monitored for a variety of reasons. There are a number of different classes into which monitoring systems could be placed. Figure 14.1 shows the general layout and interaction of these various classes.

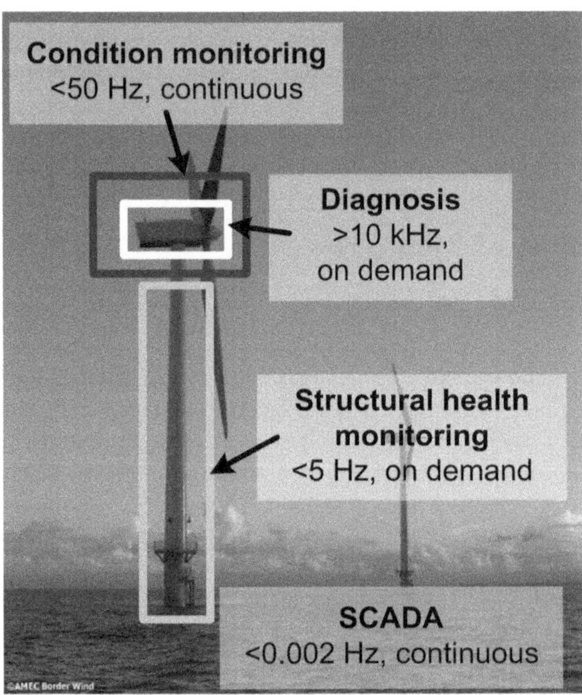

Figure 14.1   Structural health and condition monitoring of a WT

First, we have SCADA systems. Initially these systems provided measurements for WT energy production and to confirm that the WT was operational through 5–10 minute averaged values transmitted to a central database. However, SCADA systems can also provide warning of impending malfunctions in the WT drive train. The 10 minute averaged signals usually monitored in modern SCADA systems include:

- Active power output (and standard deviation over 10 minute interval)
- Anemometer-measured wind speed (and standard deviation over 10 minute interval)
- Gearbox bearing temperature
- Gearbox lubrication oil temperature
- Generator winding temperature
- Power factor
- Reactive power
- Phase currents
- Nacelle temperature (1 hour average)

This SCADA configuration is designed to show the operating condition of a WT but not necessarily give an indication of the health and a WT. However, more up-to-date SCADA systems include additional alarm settings based not only on temperature but also on vibration transducers. Often we find several vibration transducers fitted to the WT gearbox, generator bearings and the turbine main bearing. The resultant alarms are based on the level of vibration being observed over the 10 minute average period. Research is being carried out into the CM of WTs through SCADA analysis [9].

Second, there is the area of structural health monitoring (SHM). These systems aim to determine the integrity of the WT tower and foundations. SHM is generally carried out using low sampling frequencies below 5 Hz.

While SCADA and SHM monitoring are key areas for WT monitoring, this survey will concentrate on the remaining two classes of CM and diagnosis systems.

Monitoring of the drive train is often considered to be the most effective through the interaction of these two areas. CM itself may be considered as a method for determining whether a WT is operating correctly or whether a fault is present or developing. A WT operator's main interest is likely to be in obtaining reliable alarms based on CM information that can enable them to take confident action with regard to shutting down for maintenance. The operator need not know the exact nature of the fault but would be alerted to the severity of the issue by the alarm signal. Reliable CM alarms will be essential for any operator with a large number of WTs under its ownership. On this basis, CM signals need not be collected on a high frequency basis as this will reduce bandwidth for transmission and space required for storage of data.

Once a fault has been detected through a reliable alarm signal from the CMS, a diagnosis system could be activated either automatically or by a monitoring engineer to determine the exact nature and location of the fault. For diagnosis systems, data recorded at a high sampling frequency is required for analysis; however, this should only be collected on an intermittent basis. The operational time of the system should be configured to provide enough data for detailed analysis but not to flood the monitoring system or data transmission network with excess information.

Finally, Figure 14.2 gives an indication of three sections of a WT that may require monitoring based on reliability data such as that in Reference 9. While each of the three areas are shown as separate entities, CM must blur the boundaries between them in order to provide clear alarms and, subsequently, diagnostic information.

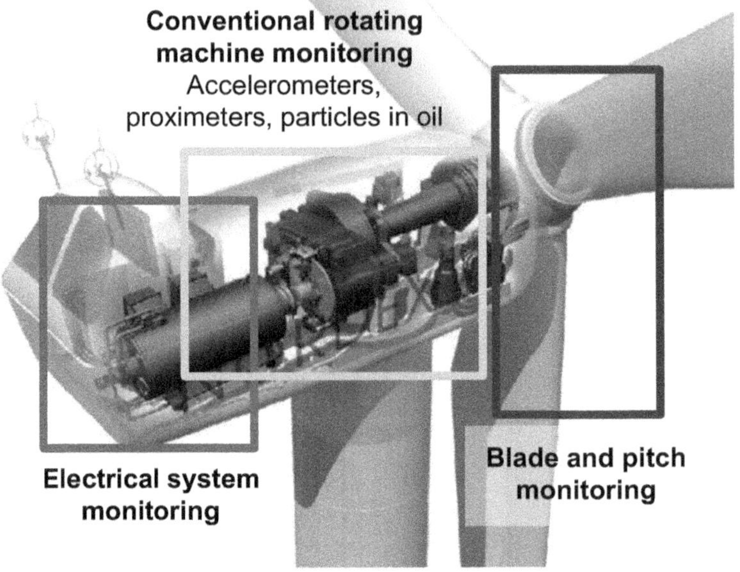

Figure 14.2   Layout of three areas for condition monitoring within the nacelle of a WT

Many of the CMSs included in this survey are a combination of CMSs and diagnostic systems due to the high level of interaction that can exist between the two types of system.

14.4   Commercially available condition monitoring systems

Table 14.1 provides a summary of a number of widely available and popular CMSs for WTs. The information has been collected from interaction with CMS and WT OEMs and product brochures over a long period of time and is up-to-date up to 2011. However, since some information has been acquired through discussion with sales and product representatives and not from published brochures, it should be noted that the table is only as accurate as the given information. The systems in Table 14.1 are arranged alphabetically by product name.

Table 14.1 Commercially available condition monitoring systems

	Product and company information			Product details (based on available literature and contact with industry including EWEC 2008, 2009, 2010, 2011)				
Ref.	Product	Supplier or manufacturer (known users)	Country of origin	Description	Main components monitored	Monitoring technology	Analysis method(s)	Data rate or sampling frequency
1	ADAPT.wind	GE Energy	USA	Up to 150 static variables monitored and trended per WT. Planetary cumulative impulse detection algorithm to detect debris particles through the gearbox planetary stage. Dynamic energy index algorithm to spread the variation over five bands of operation for spectral energy calculations and earlier fault detection. Alarm, diagnostic, analytic and reporting capabilities facilitate maintenance with actionable recommendations. Possible integration with SCADA system.	Main bearing, gearbox, generator	Vibration (accelerometer) Oil debris particle counter	FFT frequency domain analysis Time domain analysis	–
2	APPA System	OrtoSense	Denmark	Oscillation technology based on interference analysis that replicates the human ear's ability to perceive sound.	Main bearing, gearbox and generator	Vibration	Auditory perceptual pulse analysis (APPA)	–
3	Ascent	Commtest	New Zealand	System available in three complexity levels. Level 3 includes frequency band alarms, machine template creation, statistical alarming.	Main shaft, gearbox, generator	Vibration (accelerometer)	FFT frequency domain analysis Envelope analysis Time domain analysis	–
4	Brüel & Kjaer Vibro	Brüel & Kjaer (Vestas)	Denmark	Vibration and process data automatically monitored at fixed intervals and remotely sent to the diagnostic server. User-requested time waveforms for frequency and time series analysis. Time waveform	Main bearing, gearbox, generator, nacelle. Nacelle temperature.	Vibration Temperature sensor Acoustic	Time domain FFT frequency analysis	Variable up to 40 kHz. 25.6 kHz.

(Continues)

Table 14.1   (Continued)

	Product and company information			Product details (based on available literature and contact with industry including EWEC 2008, 2009, 2010, 2011)				
Ref.	Product	Supplier or manufacturer (known users)	Country of origin	Description	Main components monitored	Monitoring technology	Analysis method(s)	Data rate or sampling frequency
				automatically stored before and after user-defined event allowing advanced vibration post-analysis to identify developing faults.	Noise in the nacelle			
5	CMS	Nordex	Germany	Start-up period acquires vibration 'fingerprint' components. Actual values automatically compared by frequency, envelope and order analysis, with the reference values stored in the system. Some Nordex turbines also use the Moog Insensys fibre optic measurement system.	Main bearing, gearbox, generator	Vibration (accelerometer)	Time domain based on initial 'fingerprint'	–
6	Condition based maintenance system (CBM)	GE (Bently Nevada)	USA	This is built upon the Bently Nevada ADAPT.wind technology and System 1. Basis on System 1 gives monitoring and diagnostics of drive train parameters such as vibration and temperature. Correlate machine information with operational information such as machine speed, electrical load and wind speed. Alarms are sent via the SCADA network.	Main bearing, gearbox, generator, nacelle Optional bearing and oil temperature	Vibration (accelerometer)	FFT frequency domain analysis Acceleration enveloping	–
7	Condition based monitoring system	Bachmann Electronic GmbH	Austria	Up to nine piezoelectric acceleration sensors per module. Basic vibration analysis with seven sensors. PRÜFTECHNIK solid	Main drive train components Generator	Vibration (accelerometer)	Time domain FFT frequency analysis	24-bit res 190 kHz sample rate per channel

			borne sound sensors used for low frequency diagnostics of slowly rotating bearings on the WT LSS. Three channels for the ±10 V standard signal per module. Fully integration in Bachmann's M1 automation control system.				0.33 Hz (solid borne sound sensors)
8	Condition diagnostics system	Winergy	Denmark	Up to six inputs per module. Advanced signal processing of vibration levels, load and oil to give automated machinery health diagnostics, forecasts and recommendations for corrective action. Automatic fault identification is provided. Relevant information provided in an automated format to the operations and maintenance centre without any experts being involved. Information delivered to the appropriate parties in real time. Pitch, controller, yaw and inverter monitoring can also be included.	Main shaft, gearbox, generator	Vibration (accelerometer) Oil debris particle counter	Time domain FFT frequency domain analysis	96 kHz per channel
9	Condition management system	Moventas	Finland	Compact system measuring temperature, vibration, load, pressure, speed, oil aging and oil particle count. Sixteen analogue channels can be extended with adapter. Data acessed remotely via TCP/IP. Mobile interface available.	Gearbox, generator, rotor, turbine controller	Vibration Oil quality/ particles Torque Temperature	Time domain (possible FFT)	–
10	Distributed condition monitoring system	National Instruments	USA	Up to 32 channels; default configuration: 16 accelerometer/ microphone, 4 proximity probe and 8 tachometer input	Main bearing, gearbox, generator	Vibration Acoustic	Spectral analysis Level measurements Order analysis Waterfall plots	24-bit res 23.04 kHz of bandwidth with

(*Continues*)

Table 14.1   (*Continued*)

Product and company information			Product details (based on available literature and contact with industry including EWEC 2008, 2009, 2010, 2011)				
Ref. Product	Supplier or manufacturer (known users)	Country of origin	Description	Main components monitored	Monitoring technology	Analysis method(s)	Data rate or sampling frequency
			channels. Also provided mixed-measurement capability for strain, temperature, acoustics, voltage, current and electrical power. Oil particulate counts and fibre optic sensing can also be added to the system. Possible integration into SCADA systems.			Order tracking Shaft centre-line measurements Bode plots	antialiasing filters per accelerometer/ microphone channel
11 OneProd wind system	Areva (01dB-Metravib)	France	Eight to 32 channels. Instrumentation includes operating condition channels to trigger data acquisitions, measurement channels for surveillance and diagnosis. Data set comparison when relating to similar operating conditions; data alarm systems warn on the repetitive and abnormal shocks enabling the detection of failure modes; built-in diagnostic tool. Optional additional sensors for shaft displacement, for permanent oil quality monitoring, low frequency sensors on the structure and current and voltage sensors for generator monitoring.	Main bearing on LSS, Bearing on gearbox LSS, Bearing on intermediate gearbox shaft, on gearbox high-speed shaft, on generator Oil debris, structure, shaft displacement, electrical signals	Vibration Electrical signature analysis Thermography Oil debris particle counter	Time domain FFT frequency analysis	–
12 SMP-8C	Gamesa Eolica	Spain	Continuous on-line vibration measurement of main shaft, gearbox and generator. Comparison of spectra trends. Warnings and alarm transmission	Main shaft, gearbox, generator	Vibration	FFT frequency domain	–

13	System 1	Bently Nevada (GE)	USA	Monitoring and diagnostics of drive train parameters such as vibration and temperature. Correlate machine information with operational information such as machine speed, electrical load and wind speed.	Main bearing, gearbox, generator, nacelle Optional bearing and oil temperature	Vibration (accelerometer)	FFT frequency domain Acceleration enveloping	—
14	TCM (turbine condition monitoring) Enterprise V6 Solution with SCADA integration	Gram & Juhl A/S	Denmark	Advanced signal analysis and process signals combined with automation rules and algorithms for generating references and alarms. M-System hardware features up to 12/24 synchronous channels, interface for structural vibration monitoring and RPM sensors, external process parameters and analog outputs. TCM server stores data and does the post data processing. Control room with web based operator interface.	Tower, blades, shaft and nacelle Main bearing, gearbox and generator	Vibration (accelerometer) Sound analysis Strain analysis Process signals analysis	FFT and Wavelet frequency domain analysis Envelope analysis RMS analysis Order tracking analysis	—
15	Wind AnalytiX	ICONICS	USA	This software solution uses fault detection and diagnostics technology that identifies equipment and energy inefficiencies and provides possible causes that help in predicting plant operations, resulting in reduced downtime and costs related to diagnostic and repair.	Main WT components	Vibration (accelerometer)	Unknown	—

(Continues)

Table 14.1 (*Continued*)

	Product and company information			Product details (based on available literature and contact with industry including EWEC 2008, 2009, 2010, 2011)				
Ref.	Product	Supplier or manufacturer (known users)	Country of origin	Description	Main components monitored	Monitoring technology	Analysis method(s)	Data rate or sampling frequency
16	WindCon 3.0	SKF (REpower)	Sweden	Lubrication, blade and gearbox oil systems can be remotely monitored through SKF ProCon sofetware. WindCon 3.0 collects, analyses and compiles operating data that can be configured to suit management, operators or maintenance engineers.	Blade, main bearing, shaft, gearbox, generator, tower, generator electrical	Vibration (accelerometer, proximity probe) Oil debris particle counter	FFT frequency domain analysis Envelope analysis Time domain analysis	Analogue: DC to 40 kHz (Variable, chan dependent) Digital: 0.1 Hz– 20 kHz
17	Wind Turbine In-Service	ABS Consulting	USA	Data gathered from inspections, vibration sensors and SCADA system. Ekho for WIND software features regular diagnostics, dynamic performance reports, key performance indicators, fleet-wide analysis, forecasts/schedules and asset benchmarking. It generates alarms and notifications or triggers work orders for inspections or repairs.	Main bearing, gearbox and generator Gearbox and gear oil, rotor blades and coatings	Vibration Inspections	FFT frequency domain analysis Time domain analysis	–
18	WinTControl	Flender Service GmbH	Germany	Vibration measurements are taken when load and speed triggers are realised. Time and frequency domain analysis are possible.	Main bearing, gearbox, generator	Vibration (accelerometer)	FFT frequency domain Time domain analysis	32.5 kHz
19	WiPro	FAG Industrial Services GmbH	Germany	Measurement of vibration and other parameters given appropriate sensors. Time and frequency domain analysis carried out during alarm situations. Allows speed-dependent frequency band tracking and speed-variable alarm level.	Main bearing, shaft, gearbox, generator, temperature (Adaptable inputs)	Vibration (accelerometer)	FFT frequency domain Time domain analysis	Variable up to 50 kHz

20	WP4086	Mita-Teknik	Denmark	Up to eight accelerometers for real-time frequency and time domain analysis. Warnings/alarms set for both time and frequency domains based on predefined statistical/thresholds-based vibration limits. Operational parameters recorded alongside with vibration signals/spectra and full integration into gateway SCADA system. Algorithm toolbox for diagnostic analysis. Approximately 5000–8000 variables covering different production classes.	Main bearing, gearbox, generator	Vibration (accelerometer)	FFT amplitude spectra FFT envelope spectra Time domain magnitude Comb filtering, whitening, Kurtogram analysis	12-bit chan res Variable up to 10 kHz
21	HYDACLab	HYDAC Filtertechnik GmbH	Germany	Permanent monitoring system to monitor particles (including air bubbles) in hydraulic and lube oil systems.	Lubrication oil and cooling fluid quality	Oil debris particle counter	N/A	–
22	PCM200	Pall Industrial Manufacturing (Pall Europe Ltd)	USA (UK)	Fluid cleanliness monitor reports test data in real-time so ongoing assessments can be made. Can be permanently installed or portable.	Lubrication oil cleanliness	Oil cleanliness sensor	N/A	–
23	TechAlert 10 TechAlert 20	MACOM	UK	TechAlert 10 is an inductive sensor to count and size ferrous and non-ferrous debris in circulating oil systems. TechAlert 20 is a magnetic sensor to count ferrous particles.	Lubrication oil quality	Inductive or magnetic oil debris particle counter	N/A	–
24	BLADEcontrol	IGUS ITS GmbH	Germany	Accelerometers are bonded directly to the blades and a hub measurement unit transfers data wirelessly to the nacelle. Blades are assessed by comparing spectra with those stored for common conditions. Measurement and analysis data are stored centrally and blade condition displayed using a web browser.	Blades	Accelerometer	FFT frequency domain	= 1 kHz

(*Continues*)

Table 14.1   (*Continued*)

	Product and company information			Product details (based on available literature and contact with industry including EWEC 2008, 2009, 2010, 2011)				
Ref.	Product	Supplier or manufacturer (known users)	Country of origin	Description	Main components monitored	Monitoring technology	Analysis method(s)	Data rate or sampling frequency
25	FS2500	FibreSensing	Portugal	BraggSCOPE measurement unit designed for industrial environments to interrogate up to four Fibre Bragg Grating sensors. Acceleration, tilt, displacement, strain, temperature and pressure measurable.	Blades	Fibre optic	Unknown	Up to 2 kHz
26	RMS (rotor monitoring system)	Moog Insensys Ltd.	UK	Modular blade sensing system consisting of 18 sensors, 6 per blade, installed in the cylindrical root section of each blade to provide edgewise and flapwise bending moment data. Can be designed-in during turbine manufacture or retrofitted. Monitors turbine rotor performance, mass and aerodynamic imbalance, blade bending moments, icing, damage and lightning strikes. Possible integration, as an external input, in commercial available CMSs.	Blades	Fibre optic strain	Time domain strain analysis	25 Hz/sensor

The first observation to make is that the CMSs nearly all focus on the same WT sub-assemblies as follows:

- Blades
- Main bearing
- Gearbox internals
- Gearbox bearings
- Generator bearings

Summarising Table 14.1 using the numbers there shows that there are:

- 20 systems primarily based on drive train vibration analysis (1–20),
- 3 systems solely for oil debris monitoring (21–23),
- 1 system using vibration analysis for WT blade monitoring (24),
- 2 systems based on fibre optic strain measurement in WT blades (25 and 26).

The majority of the systems are based on monitoring methods originating from other, traditional rotating machinery industries. Indeed 19 of the 26 systems in the table are based on vibration monitoring using accelerometers typically using a configuration similar to that shown in Figure 14.3 for the Mita-Teknik WP4086 CMS (20).

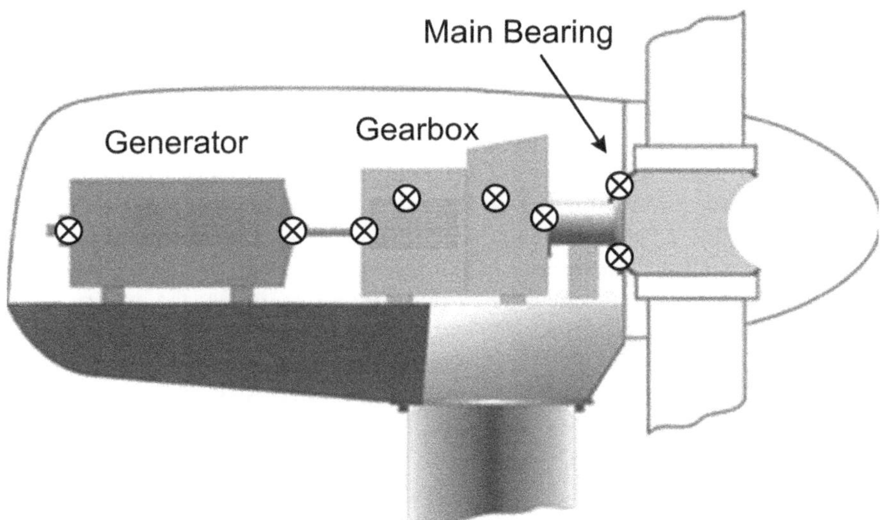

*Figure 14.3   Typical CMS accelerometer positions in the nacelle of a WT
[Source: [10]]*

Of these 19 CMSs, all have the capability to carry out some form of diagnostic procedure once a fault has been detected. In most cases this is done through fast Fourier transform (FFT) analysis of high frequency data in order to detect fault-specific frequencies. In the case of the SKF WindCon 3.0 (16), the Areva OneProd Wind CMS (11) and several others, high data acquisition is triggered by operational parameters. For example, the SKF WindCon 3.0 CMS can be configured to collect a vibration spectrum on either a time basis or when a specific load and speed condition is achieved. The aim of this is to acquire data that is directly comparable between each point and, importantly, to allow spectra to be recorded in apparently stationary conditions. This is important to note when using traditional signal processing methods such as the FFT that require stationary signals in order to obtain a clear result. The Mita-Teknik WP4086 system (20), however, states that it includes advanced signal processing techniques such as comb filtering, whitening and Kurtogram analysis that in combination with re-sampling and order alignment approaches allow the system to overcome the effects of WT speed variations.

An innovative vibration-based CMS is OrtoSense APPA (2), which is based on auditory perceptual pulse analysis. This patented technology outperforms the human ear by capturing a detailed interference pattern and detecting even the smallest indication of damaged or worn elements within the machine/turbine. OrtoSense states that its product is four to ten times more sensitive compared to prevailing systems.

Five of the vibration-based CMSs also state that they are able to monitor the level of debris particles in the WT gearbox lubrication oil system. Further to this, included in the table are three systems that are not in themselves CMSs. These three (21–23) are oil quality monitoring systems or transducers rather than full CMSs but are included, as discussion with industry has suggested that debris in oil plays a significant role in the damage and failure of gearbox components. Systems using this debris in oil transducers are using either cumulative particle counts or particle count rates.

Several of the 20 vibration-based CMSs also allow for other parameters to be recorded alongside vibration, such as load, wind speed, generator speed and temperatures, although the capabilities of some systems are unclear given the information available. There is some interest being shown as regards the importance of operational parameters in WT CM. This arises from the fact that many analysis techniques, for example the FFT, have been developed in constant speed, constant load environments. This can lead to difficulties when moving to the variable speed, variable load WT; however, experienced CM engineers are able to use these techniques and successfully detect faults.

Recent CM solutions, such as (1), (10), (14), (20), can be adapted and fully integrated with existing SCADA systems using standard protocols. Thanks to this integration, the analysis of the systems installed on the wind energy plant can also directly consider any other signals or variables of the entire controller network, as for example current performance and operating condition, without requiring a doubling of the sensor system. The database, integrated into a single unified plant operations' view, allows a trend analysis of the condition of the machine.

In some cases the CMS company offers also custom service solutions from 24/7 remote monitoring to on-demand technical support, examples are GE Energy ADAPT.wind (1), ABS Wind Turbine In-Service (17) and several others.

Two in the table (25 and 26) are effectively Balde monitoring systems (BMS) based on strain measurement using fibre-optic transducers. These are aimed at detection of damage to WT blades and, in the case of the Moog Insensys system (26), blade icing, mass unbalance or lightning strikes. Both systems may be fitted to WT blades retrospectively. Compared to vibration monitoring techniques, these systems can be operated at low sampling rates as they are looking to observe changes in time domain. They are usually integrated in the WT control system but there are also some cases of integration as an external input into commercially available conventional vibration-based CMSs. In addition to (25) and (26) there is the IGUS BMS (24) that uses accelerometers to monitor blade damage, icing and lightning strikes. This system compares the blade accelerometer FFT with stored spectra for similar operating conditions and has the power to automatically shut down or restart a WT based on the results. The system appears to be popular within the industry.

14.5   Future of wind turbine condition monitoring

As can be seen from this survey of current CMSs there is a clear trend towards vibration monitoring of WTs. This is presumably a result of the wealth of knowledge gained from many years work in other fields. It is likely that this trend will continue; however, it would be reasonable to assume that other CM and diagnostic techniques will be incorporated into existing systems.

Currently, these additions are those such as oil debris monitoring and fibre optic strain measurement. However, it is likely that major innovation will occur in terms of developing signal processing techniques. In particular, the industry is already noting the importance of operational parameters such as load and speed and so techniques may begin to adapt further to the WT environment leading to more reliable CMSs, diagnostics and alarm signals.

Automation of CM and diagnostic systems may also be an important development as WT operators acquire a larger number of turbines and manual inspection of data becomes impractical. Further to this, it is therefore essential that methods for reliable, automatic diagnosis are developed with consideration of multiple signals in order to improve detection and increase operator confidence in alarm signals.

However, it should be noted that a major hindrance to the development of CMSs and diagnostic techniques could be data confidentiality, which means that few operators are able to divulge or obtain information concerning their own WTs. This is an issue that should be addressed if the art of CM is to progress quickly. Confidentiality has also led to a lack of publicly available cost justification of WT CM, which seems likely to provide overwhelming support for WT CM, particularly in the offshore environment where availability is at a premium.

14.6   Summary

From this survey we can conclude that:

- There is a wide variety of commercially available CMSs currently in use on operational WTs.
- Monitoring technology is currently based on techniques from other, conventional rotating machine industries.
- Successful CMSs must be able to adapt to the non-stationary, variable speed nature of WTs.
- Vibration monitoring is currently favoured in commercially available systems using standard time and frequency domain techniques for analysis.
- These traditional techniques can be applied to detect WT faults but require experienced CM engineers for successful data analysis and diagnosis.
- Some commercially available CMSs are beginning to adapt to the WT environment and to be fully integrated into existing SCADA systems.
- A diverse range of new or developing technologies are moving into the WT CM market.

Finally, it should be noted that there is no current consensus in the WT industry as to the correct route forward for WT CMS. Work in this document and its references suggest that CM of WTs will be important for large onshore WTs, essential for all offshore development and should be considered carefully by the industry as a whole.

14.7   References

[1]   Windstats (WSD & WSDK) quarterly newsletter, part of WindPower Weekly, Denmark. Available from www.windstats.com [Last accessed 8 February 2010]

[2]   Landwirtschaftskammer (LWK). Schleswig-Holstein, Germany. Available from http://www.lwksh.de/cms/index.php?id¼1743 [Last accessed 8 February 2010]

[3]   Noordzee Wind Various Authors:
   a. 'Operations Report 2007', Document No. OWEZ_R_000_20081023, October 2008. Available from http://www.noordzeewind.nl/files/Common/Data/OWEZ_R_000_20081023%20Operations%202007.pdf?t=1225374339 [Accessed January 2012]
   b. 'Operations Report 2008', Document No. OWEZ_R_000_ 200900807, August 2009. Available from http://www.noordzeewind.nl/files/Common/Data/OWEZ_R_000_20090807%20Operations%202008.pdf [Accessed January 2012]
   c. 'Operations Report 2009', Document No. OWEZ_R_000_20101112, November 2010. Available from http://www.noordzeewind.nl/files/Common/Data/OWEZ_R_000_20101112_Operations_2009.pdf [Accessed January 2012]

[4]   ReliaWind. Available from http://www.reliawind.eu [Last accessed 8 February 2010]

[5]   Tavner P.J., Faulstich S., Hahn B., van Bussel G.J.W. 'Reliability and availability of wind turbine electrical and electronic components'. *European Journal of Power Electronics*. 2011;**20**(4):45–50

[6]   Germanischer Lloyd. *Guideline for the Certification of Condition Monitoring Systems for Wind Turbines*. Edition. Hamburg, Germany: Germanischer Lloyd; 2007

[7]   Germanischer Lloyd. *Guideline for the Certification of Wind Turbines*. Edition 2003 with Supplement 2004, Reprint. Hamburg, Germany: Germanischer Lloyd; 2007

[8]   Germanischer Lloyd. *Guideline for the Certification of Offshore Wind Turbines*. Edition 2005, Reprint. Hamburg, Germany: Germanischer Lloyd; 2007

[9]   Crabtree C.J., Feng Y., Tavner P.J. 'Detecting incipient wind turbine gearbox failure: A signal analysis method for online condition monitoring'. *Proceedings of European Wind Energy Conference, EWEC2010*. Warsaw: European Wind Energy Association; 2010

[10] Isko V., Mykhaylyshyn V., Moroz I., Ivanchenko O., Rasmussen P. 'Remote wind turbine generator condition monitoring with WP4086 system'. *Proceedings of European Wind Energy Conference, EWEC2010*. Warsaw: European Wind Energy Association; 2010

Chapter 15

Appendix 6: Weather, its influence on offshore wind reliability

15.1 Wind, weather and large WTs

15.1.1 Introduction

Weather conditions are difficult to describe succinctly for engineering purposes and it is not yet clear which aspects are important for WT operation. But the weather has been measured at sea since 1805, using the Beaufort scale summarised in Table 15.1, and this is a helpful basis for understanding the impact of weather on offshore wind farms.

It is important to appreciate from Table 15.1 the very large range of weather conditions to which large WTs, remote unmanned robotic power units operating 24/7, are exposed and operate successfully.

Compare this to the relatively benign environmental conditions prevailing in conventional fossil- and nuclear-fired or hydro power stations.

The impact of the sea and wind conditions, over the ranges shown in Table 15.1, on the foundations, base, tower, nacelle and operational components of operating offshore WTs need to be borne in mind by all parts of the wind industry, especially with regard to O&M. It should be particularly noted that the wind speed and wave heights or sea condition, shown on the Beaufort scale, are not necessarily contemporaneous because wind speeds may be rising before a storm when wave heights are not fully established, while after the storm large wave swell may still persist when wind speeds have moderated.

Weather conditions are studied further under the following headings that are considered important for offshore wind farms at the current time.

15.1.2 Wind speed

The range of WT operational wind speeds, with cut-in at 2–3 m/s and cut-out at 26 m/s, is highlighted in Table 15.1 in light grey. This ranges from Beaufort Force 2 to 9, that is from light breezes to strong gale. Furthermore some WTs, notably the Enercon large WT range, utilise storm control and do not cut-out sharply at 26 m/s but rather adopt a reducing power control from full power at 28 m/s to zero power at 34 m/s, extending their operation to just above Force 10, violent storm.

Table 15.1   The Beaufort scale

Beaufort number	Description	Wind speed	Wave height	Sea conditions
0	Calm	<1 km/h <0.3 m/s <1 mph <1 knot	0 m 0 ft	Flat
1	Light air	1.1–5.5 km/h 0.3–2 m/s 1–3 mph 1–2 knot	0–0.2 m 0–1 ft	Ripples without crests
2	Light breeze	5.6–11 km/h 2–3 m/s 4–7 mph 3–6 knot	0.2–0.5 m 1–2 ft	Small wavelets; crests of glassy appearance, not breaking
3	Gentle breeze	12–19 km/h 3–5 m/s 8–12 mph 7–10 knot	0.5–1 m 2–3.5 ft	Large wavelets; crests begin to break; scattered whitecaps
4	Moderate breeze	20–28 km/h 6–8 m/s 13–17 mph 11–15 knot	1–2 m 3.5–6 ft	Small waves with breaking crests; fairly frequent whitecaps
5	Fresh breeze	29–38 km/h 8.1–10.6 m/s 18–24 mph 16–20 knot	2–3 m 6–9 ft	Moderate waves of some length; many whitecaps; small amounts of spray
6	Strong breeze	39–49 km/h 10.8–13.6 m/s 25–30 mph 21–26 knot	3–4 m 9–13 ft	Long waves begin to form; white foam crests are very frequent; some airborne spray is present
7	High wind, moderate gale, near gale	50–61 km/h 13.9–16.9 m/s 31–38 mph 27–33 knot	4–5.5 m 13–19 ft	Sea heaps up; some foam from breaking waves is blown into streaks along wind direction; moderate amounts of airborne spray
8	Gale, fresh gale	62–74 km/h 17.2–20.6 m/s 39–46 mph 34–40 knot	5.5–7.5 m 18–25 ft	Moderately high waves with breaking crests forming spindrift; well-marked streaks of foam are blown along wind direction; considerable airborne spray
9	Strong gale	75–88 km/h 20.8–24.4 m/s 47–54 mph 41–47 knot	7–10 m 23–32 ft	High waves whose crests sometimes roll over; dense foam is blown along wind direction; large amounts of airborne spray may begin to reduce visibility

(Continues)

Table 15.1   (*Continued*)

Beaufort number	Description	Wind speed	Wave height	Sea conditions
10	Storm, whole gale	89–102 km/h 24.7–28.3 m/s 55–63 mph 48–55 knot	9–12.5 m 29–41 ft	Very high waves with overhanging crests; large patches of foam from wave crests give the sea a white appearance; considerable tumbling of waves with heavy impact; large amounts of airborne spray reduce visibility
11	Violent storm	103–117 km/h 28.6–32.5 m/s 64–72 mph 56–63 knot	11.5–16 m 37–52 ft	Exceptionally high waves; very large patches of foam, driven before the wind, cover much of the sea surface; very large amounts of airborne spray severely reduce visibility
12	Hurricane force	118 km/h (32.8 m/s)	14 m	Huge waves; sea is completely white with foam and spray; air is filled with driving spray, greatly reducing visibility

From force 3 to 9 is the operating range of WTs; From wave height 0 to 2 m is the operating range of smaller access vessels to offshore WTs. *[Source: [1]]*

From the work in Chapter 3, Figure 3.2, the author first noticed that high WT failure rates were related to high wind speed, particularly in the stormy weather in the winters of 1998 and 1999 in Denmark and Germany. This was studied particularly across Denmark [2] and in a later more precise study of three specific wind farms in Germany [3], the reliability results from which are presented in Section 15.3.1, 15.3.2, 15.3.3 and 15.3.4. These three wind farms in Germany were located at:

- Fehmarn, located on a small island in the Baltic Sea, off the coast of Schleswig Holstein, Germany
- Krummhörn, located on the North Sea Coast in Lower Saxony, Germany
- Ormont, located inland in the highlands in Rhineland Palatinate, Germany

An indication of the annual variation of wind speed at these three disparate sites is shown in Figure 15.1.

15.1.3   *Wind turbulence*

Wind speed has a significant influence on reliability but probably more important to WT reliability is the wind turbulence.

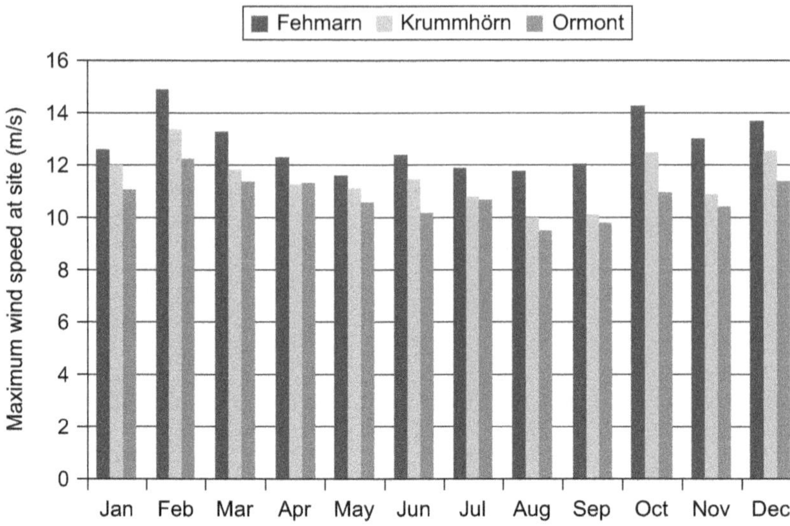

*Figure 15.1   Annual variation in horizontal wind speeds at three disparate
German onshore wind farms [Source: [3]]*

Wind turbulence refers to wind speed fluctuations on a short timescale. However, there is no established time period over which such wind speed variations are officially classed as turbulent. Indeed as Reference 4 explains '[although] turbulence ... has been studied for over a century ... it is surprisingly difficult to define precisely what we mean by turbulence'. Turbulent eddies are formed in the atmosphere due to thermal gradients or when the wind flow passes over a rough surface or is disrupted by obstacles such as trees, hills and buildings. Wake effects from neighbouring WTs can also significantly contribute to the turbulence experienced by a WT. Due to a lack of obstacles and a relatively smooth surface for the wind to pass over, offshore sites typically experience less turbulence than onshore installations, although this does depend on the above-sea temperature gradient and sea state.

Although Reference 4 acknowledges the difficulty in defining turbulence, it does attempt to give an idea of the size of the turbulent eddies formed in the wind. It states that the largest eddies have scales of approximately 100 m and that the smallest are approximately 1 mm. This translates to most turbulent variations lasting less than 100 seconds at a fixed position.

Power spectral analysis shows the wind speed variation timescales containing the most energy. The defining work in this area by van der Hoeven [5] identified a 'turbulent peak', as shown in Figure 15.2, at a period of about 1 minute for horizontal wind at a height of 100 m. It is probable that the eddies which have the most effect upon the WT and its drive train, known to be vulnerable to the fatigue caused by turbulence, see Chapter 3, are those with dimensions significant compared to the blade length or disc diameter, i.e. 25–125 m for large WTs.

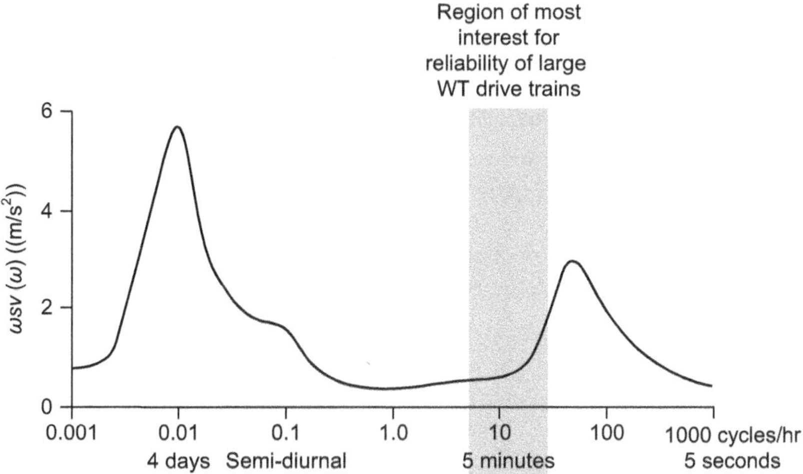

*Figure 15.2   Van der Hoeven power spectrum of horizontal wind speeds
[Source: [1]]*

The IEC standard [6] uses a measure called the turbulence intensity, *I*, which is the ratio of the wind speed standard deviation, σ, to the mean wind speed, *u*, for each 10 minute reporting period. This is the measure of turbulence used throughout the wind energy industry; both *I* and *u* are readily available from most WT SCADA systems and met masts.

$$I = \frac{\sigma}{u} \tag{15.1}$$

Due to the definition of this measure, when *u* is small, *I* becomes large, but is physically insignificant. Therefore, some advocate that *I* for wind speeds below 8–10 m/s are not load relevant.

When describing *I* over a period of time greater than 10 minutes, it is necessary to perform some kind of average using the 10 minute values. A number of subtly different terms are used in the industry. Characteristic turbulence intensity, I_{char}, is used in the second edition of the standard [6]; representative turbulence intensity, I_{rep}, is used in the third edition. They differ in that I_{char} is defined as the mean plus the standard deviation, whereas I_{rep} is the 90% percentile. Both I_{char} and I_{rep} are used in the wind industry, but I_{char} is more popular and will be used for cross-correlations below.

There is also ambiguity over the term 'gust'. Short-term, extreme events would fit with the term's usage in IEC 61400-1 [6], which provides an extreme operating gust (EOG) model that simulates a rapid wind speed increase, for example 24 36 m/s over 5 seconds. For the purposes of this section gusts will be assumed to be special cases within a wind velocity spectrum, which may be considered to be short-term, extreme event forms of turbulence.

15.1.4   Wave height and sea condition

The range of wave heights and sea conditions for which smaller access vessels, such as the vessel shown in Figure 8.2, can transfer personnel safely in an offshore wind farm is highlighted in darker grey in Table 15.1. It ranges from Beaufort Force 0 to 4, that is up to a moderate breeze. There is as yet no information on the systematic effects of sea state on offshore WT reliability

15.1.5   Temperature

The annual variation of temperature at the three onshore German sites described above is shown in Figure 15.3. It should be noted that the island location, Fehmarn, has the least temperature variance, as one would expect and this is what we may expect at offshore locations.

15.1.6   Humidity

Similarly, the variation of humidity at these three onshore German sites is shown in Figure 15.4. Again it should be noted that it is the coastal location, Krummhörn, which has the highest mean humidity and least humidity variance and this is what we may expect at offshore locations.

15.2   Mathematics to analyse weather influence

15.2.1   General

References 2 and 3 used failure time periodograms and failure-environmental cross-correlograms to analyse the influence of weather phenomena on failures.

15.2.2   Periodograms

This approach transforms the time domain data into the frequency domain using Fourier analysis. If a signal, $f(t)$, is periodic, that is

$$f(t) = f(t + T) \tag{15.2}$$

then it is possible to represent it in the frequency domain. This may be restated as

$$F(s_k) = \frac{1}{T} \int_{-T/2}^{T/2} f(t)e^{-j2\pi s_k t} dt \tag{15.3}$$

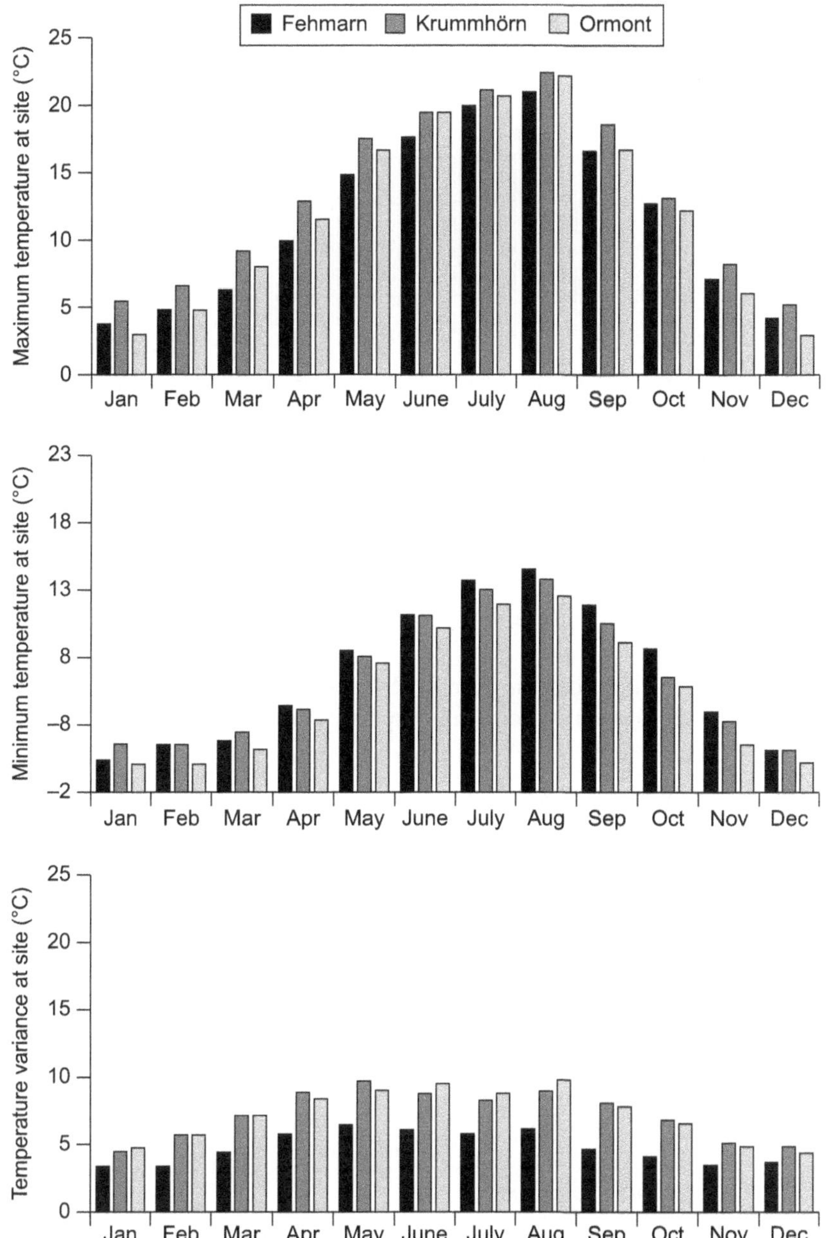

Figure 15.3   Annual variation of temperatures at three disparate German onshore wind farms [Source: [3]]

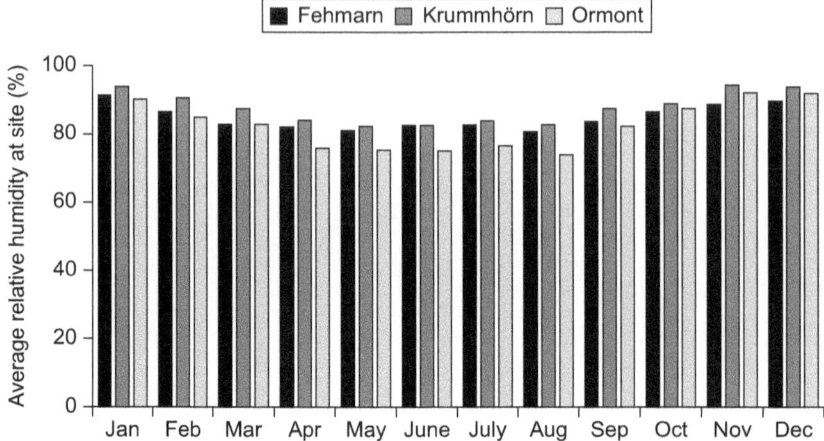

*Figure 15.4   Annual variation of humidity at three disparate German onshore
wind farms [Source: [3]]*

where $k = 0, \pm1, \pm2, \ldots$ and denotes the kth-harmonic of the fundamental frequency $(1/T)$.

In this case the time domain data is sampled, so the transformation from the time domain to the frequency domain is expressed by

$$F(s_k) = \frac{1}{N}\sum_{n=0}^{N-1}[f(t_n)]e^{-j2\pi nk/N} \tag{15.4}$$

The transformation was computed using the FFT, which is a well established and computationally efficient way of obtaining a discrete Fourier transform (FT). It is only strictly valid to carry out an FT on a periodic signal. When taking the signal FT it is assumed that the fundamental is the reciprocal of the signal length. If this requirement is not met there will be a discontinuity in the signal, resulting in harmonic leakage in the frequency domain. For the present purposes this assumption is unlikely to be valid; therefore, it was necessary to apply a Hanning window to minimise the harmonic leakage.

15.2.3   Cross-correlograms

Cross-correlation is a time domain technique used to measure the extent to which two signals are linearly related. The cross-correlation function of two stationary time domain signals, $f(t)$ and $g(t)$, is given by

$$R_{fg}(\tau) = \int_{-\infty}^{\infty} f(t)g(t+\tau)dt \tag{15.5}$$

This may be restated as

$$R_{fg}(\tau) = \lim_{T \to \infty} \frac{1}{2T} \int_{-T}^{T} f(t)g(t + \tau)dt \qquad (15.6)$$

where T is the period of observation, that is the signal length, τ is the time lag between the signals. For sampled signals this is written as

$$R_{fg}[m] = \lim_{N \to \infty} \frac{1}{2N + 1} \sum_{-N}^{N} f[n]g[n + m] \qquad (15.7)$$

where N is the number of data points and m is the lag. Note that in order to interpret this lag as a time shift the time series must be uniformly sampled.

The cross-correlation function can now be estimated where the signals $f(t)$ and $g(t)$ are of finite length. For sampled signals the biased cross-correlation is computed by

$$R_{fg}[m] = \frac{1}{N} \sum_{n=1}^{N-m+1} f[n]g[n + m] \qquad (15.8)$$

While the unbiased form is

$$R_{fg}[m] = \frac{1}{N - |m|} \sum_{n=1}^{N-m+1} f[n]g[n + m] \qquad (15.9)$$

where $m = 1,..., M + 1$.

15.2.4   Concerns

A serious concern about these analyses has been the relative frequencies of the failure and meteorological data. Failure data is usually collected daily or weekly and meteorological data may be collected at 1 minute intervals. This immediately creates a problem when trying to cross-correlate such widely disparate frequencies.

The physical reality is that the WT failure mechanisms (Figure 3.2) are essentially cumulative or integrative and this needs to be considered in the analysis methods.

15.3   Relationships between weather and failure rate

15.3.1   Wind speed

Based upon the analysis methods described above, the effect of wind speed on WT failure rates was investigated in Denmark. The detailed failure and weather data of Denmark were available from Reference 2. This showed a significant correlation

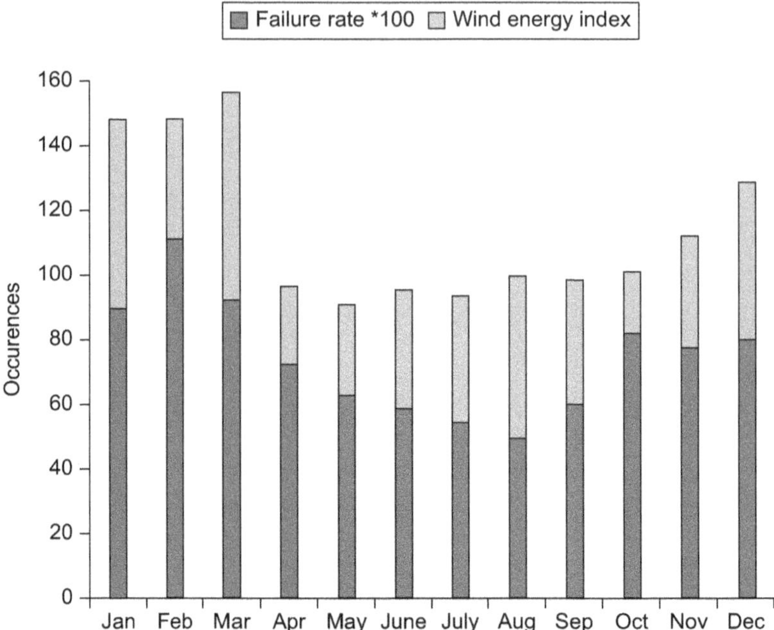

*Figure 15.5   Average onshore Danish WT monthly failure rates and WEI, related
to wind speed, for each of the 12 Months, 1994–2004 [Source: [2]]*

between WT failure rate and months when the average wind speed was higher, with a cross-correlation factor of 44%. This is shown graphically in Figure 15.5, where Danish WT failure rates during the 1994–2004 period peaked each year in February and October, while wind speeds peaked in February.

Even more illuminating was that the cross-correlation of failures in different sub-assemblies with wind speed varied across those sub-assemblies, as shown in Figure 15.6.

Surprisingly in this survey the generator proved the most sensitive sub-assembly to higher wind speeds rather than the aerodynamic sub-assemblies, for example the yaw or control systems. The pragmatic explanation for this behaviour could be that the generators for these WTs were commercially procured, standard sub-assemblies not necessarily hardened for the wind industry environment.

The strength of this study was the large number of turbines considered over an extensive period of time and the large number of failures involved. The weakness of the study was that it blurred the reliability of many different turbine designs and considered a monthly average wind speed over the whole of Denmark, thereby concealing, by averaging, more detailed wind speed effects.

An improved study [3] of German data has been prepared, which concentrates on the failure rate of one particular WT type located at onshore wind

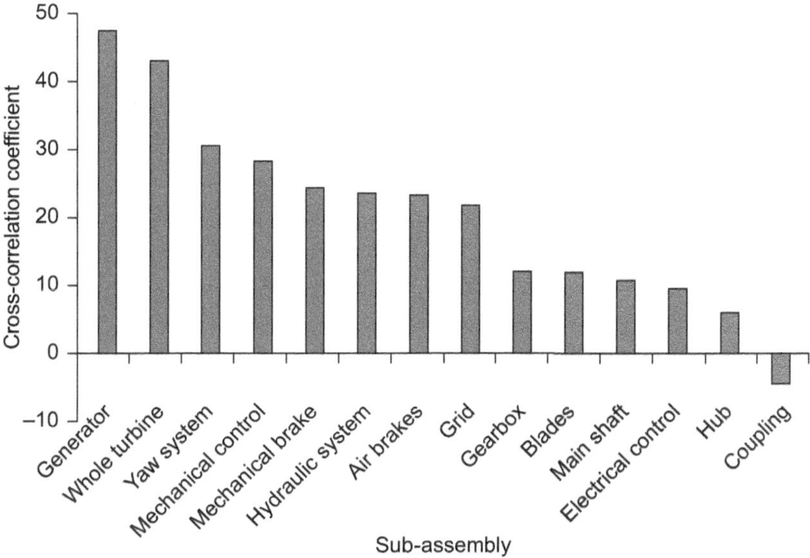

Figure 15.6   Summary of cross-correlograms of Danish onshore WT sub-assembly failure rates to wind speed, 1994–2004 [Source: [2]]

farms in locations where accurate weather data were available. The authors identified three locations, as described in Section 15.1.2, with different climatic and operating conditions, operating the same type of turbine. By taking a more focussed approach this paper corrects some of the shortcomings of the previous study of Danish failure data and reveals more significant effects of weather and location on reliability. However, it does not show a cross-correlation of failures with wind speed but does show cross-correlations with other weather factors as described below.

15.3.2   Temperature

The results of Reference 3 show an interesting consequence of temperature on WT failures, visible in Figure 15.7.

This shows that at all three sites the failure rates were affected by temperature, all showing annual variations with the seasons but with a cross-correlation phase variation between them and the general effect of temperature on failure rate being higher in summer than winter.

A more detailed observation was also shown at one site by separating failures between electrical and mechanical sub-assemblies as shown in Figure 15.8.

Figure 15.8 shows that the cross-correlation between temperature and failures is dominated by the electrical rather than mechanical sub-assemblies. The use of sealed, environmentally controlled nacelles offshore will counter this issue.

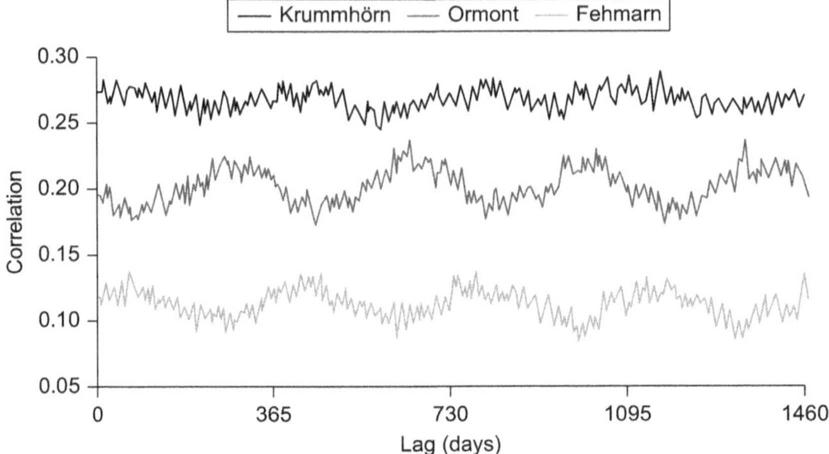

Figure 15.7   Wrapped correlograms of failure variation at three onshore wind farms due to daily maximum temperature [Source: [3]]

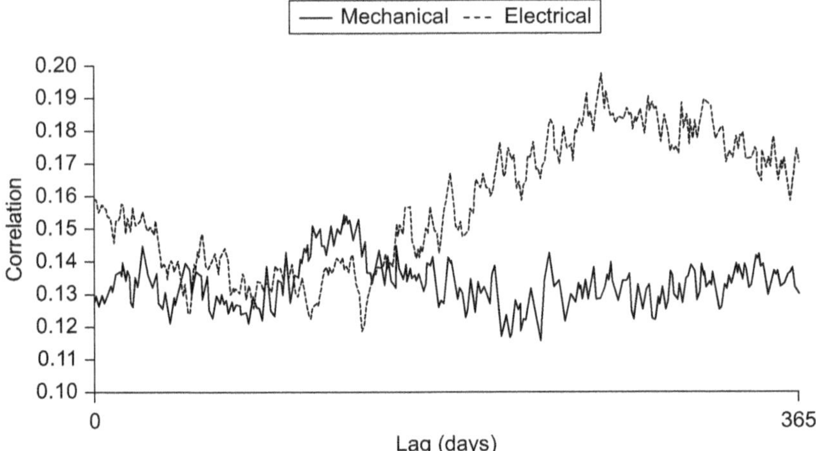

Figure 15.8   Wrapped correlograms of failure variation at Fehmarn due to daily maximum temperatures, split between electrical and mechanical failures [Source: [3]]

15.3.3   Humidity

The results of Reference 3 showed that there was a significant cross-correlation, 23–31%, with failures at the island and coastal site and a much lower cross-correlation at the inland site, suggesting that at offshore locations the issue of humidity is significant and again this is being handled by the adoption of sealed, environmentally controlled nacelles.

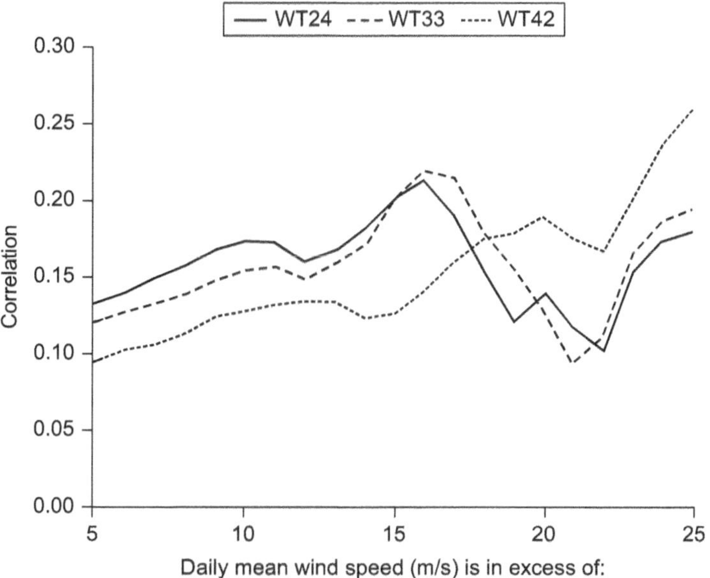

Figure 15.9   For three WTs correlograms of pitch failure with I_{char} on days when the daily mean wind speed exceeded figures shown

15.3.4   Wind turbulence

Analysis performed on SCADA data from 3 off 2 MW WTs with hydraulic pitch system faults, Figure 15.9, shows a significant correlation between failures and turbulence, represented by I_{char}, measured when the daily mean wind speed exceeded the wind speeds shown. This indicates that turbulence is driving pitch failures in this case, although the results are difficult to interpret.

Another analysis on SCADA data from 6 off 1.6 MW WTs in three wind farms with electric pitch system faults, considering a different measure of turbulence to I_{char}, again shows a significant correlation between failures and turbulence, see Table 15.2, using a turbulence measure related to wind speed, ku_2, ku_5, ku_8, ku_{10}, compared to the correlation with wind speed itself. First, the analysis shows clearly that failures are more sensitive to turbulence than wind speed; second it shows a difference between different wind farms, mentioned by the operator, wind farm 2 being known to experience more turbulent conditions and being prone to pitch failures.

A further analysis on these WTs considered the effect of gusts upon the pitch failures and does not show any particular sensitivity for wind farm 2 but rather confirms the correlations of Table 15.2.

These three sets of results demonstrate that it is possible to detect sensitivity to wind speed, turbulence and gusts of WT pitch mechanisms failures, which are known to be significant, see Table 3.1. More work needs to be done to establish the

Table 15.2   Cross-correlation of pitch failures with turbulence on 1.6 MW WTs

Variables		WT1	WT2	WT3	WT4	WT5	WT6
			WF1			WF2	WF3
Mean wind speed	u	0.15	0.24	0.23	0.15	0.11	0.18
Wind speed turbulence	ku_2	0.20	0.41	0.33	0.17	0.17	0.36
coefficients	ku_5	0.23	0.41	0.28	0.21	0.28	0.33
	ku_8	0.32	0.35	0.27	0.40	0.34	0.30
	ku_{10}	0.34	0.31	0.28	0.45	0.50	0.24

Table 15.3   Cross-correlation of pitch failures with gusts on 1.6 MW WTs

Variables	WT1	WT2	WT3	WT4	WT5	WT6
		WF1			WF2	WF3
Wind gusts over 2 m/s	0.28	0.24	0.23	0.23	0.18	0.19
Wind gusts over 5 m/s	0.33	0.47	0.37	0.31	0.31	0.34
Wind gusts over 10 m/s	0.30	0.33	0.33	0.25	0.37	0.71

link but operators need to realise that measurements made from SCADA can unlock these root causes.

15.4   Value of this information

15.4.1   To wind turbine design

Establishing the influence on WT reliability of weather conditions is at a very early stage of development in terms of both the data available and the analysis methods needed. However, the above work has shown clearly that high wind speeds, wind turbulence, gusts, temperature variance and humidity all affect reliability and they affect different sub-assemblies in different ways. The following can be concluded:

- High wind speeds, turbulence and gusts deteriorate WT blade, pitch and mechanical drive train reliability.
- Temperature and humidity variances deteriorate electrical more than mechanical sub-assembly reliabilities.
- Sealed, environmentally controlled nacelles are essential for offshore WTs.

There is very little information to date on the systematic effects of icing [7] and sea state on offshore WT reliability.

15.4.2   To wind farm operation

The above results suggest that high wind speeds and turbulence are likely to affect wind farm availability. WTs further back in an array are also likely to be less reliable than those near the leading edge due to effects of wind turbulence on them. It is again not yet clear what the effects of icing and sea state will be upon the wind farm. The issues of icing and sea state should be a significant area for further investigation for WTs and wind farms if we are to improve offshore wind reliability.

15.5 References

[1]   Wikipedia, *Beaufort scale*. Available from http://en.wikipedia.org/wiki/ Beaufort_scale [Accessed January 2012]

[2]   Tavner P.J., Edwards C., Brinkman A., Spinato F. 'Influence of wind speed on wind turbine reliability'. *Wind Engineering*. 2006;**30**(1):55–72

[3]   Tavner P.J., Greenwood D.M., Whittle M.W.G., Gindele R., Faulstich S., Hahn B.'Study of weather and location effects on wind turbine failure rates'. *Wind Energy*. (2012); in press. DOI: 10.1002/we.538, Early View.

[4]   McIlveen R. *Fundamentals of Weather and Climate*. 2nd edn. London: Chapman & Hall; 1992.

[5]   van der Hoeven I. 'Power spectrum of horizontal wind speed in the frequency range from 0.0007 to 900 cycles per hour'. *American Meteorological Society*. 1957;**14**(2):160–4.

[6]   IEC 61400-1:2005 Wind turbines – Part 1: Design Requirements. Geneva, Switzerland: International Electrotechnical Commission.

[7]   Hochart C., Fortin G., Perron J., Ilinca A. 'Wind turbine performance under icing conditions'. *Wind Energy*. 2007;**11**(4):319–33.